Studies in Space Policy

Volume 18

Edited by the European Space Policy Institute
Director: Jean-Jacques Tortora

The use of outer space is of growing strategic and technological relevance. The development of robotic exploration to distant planets and bodies across the solar system, as well as pioneering human space exploration in earth orbit and of the moon, paved the way for ambitious long-term space exploration. Today, space exploration goes far beyond a merely technological endeavour, as its further development will have a tremendous social, cultural and economic impact. Space activities are entering an era in which contributions of the humanities—history, philosophy, anthropology—, the arts, and the social sciences—political science, economics, law—will become crucial for the future of space exploration. Space policy thus will gain in visibility and relevance. The series Studies in Space Policy shall become the European reference compilation edited by the leading institute in the field, the European Space Policy Institute. It will contain both monographs and collections dealing with their subjects in a transdisciplinary way.

More information about this series at http://www.springer.com/series/8167

Marco Aliberti · Ksenia Lisitsyna

Russia's Posture in Space

Prospects for Europe

European Space Policy Institute

Marco Aliberti
European Space Policy Institute
Vienna, Austria

Ksenia Lisitsyna
European Space Policy Institute
Vienna, Austria

ISSN 1868-5307   ISSN 1868-5315   (electronic)
Studies in Space Policy
ISBN 978-3-030-08041-9   ISBN 978-3-319-90554-9   (eBook)
https://doi.org/10.1007/978-3-319-90554-9

Printed on acid-free paper

This Springer imprint is published by the registered company Springer International Publishing AG part of Springer Nature
The registered company address is: Gewerbestrasse 11, 6330 Cham, Switzerland

Acknowledgements

The authors would like to express their sincerest gratitude to the many people who have supported and contributed to completion of the present study. As a follow-up of the research activities on the Russian space programme conducted at the European Space Policy Institute (ESPI) by Charlotte Mathieu, the greatest debt is owed to her. Much appreciation is also paid to the Director of ESPI, Jean Jacques Tortora, and ESPI Coordinator of Studies, Sebastien Moranta, for their proactive inputs and useful critiques during the research and writing process of this project. Sincere gratefulness is more broadly extended to the ESPI team, in particular ESPI Resident Fellows Matteo Tugnoli and Cenan Al-Ekabi and ESPI Research Intern Leyton Wells, for their relevant feedback and cooperation in the finalisation of the study.

The study has immensely benefited from the openness and wisdom of many interlocutors external to ESPI. The authors are indebted to all the experts and stakeholders interviewed under Chatham House Rule for this report, who contributed with their experience and outstanding insights to the value, depth and balance of this study. A specific mention goes to Igor Smirnov, Alexander Serkin, Alexander Shaenko, Dmitry Payson, Pavel Pushkin, Vitaly Egorov, Alexander Khokhlov, Frederic Nordlund, Karl Bergquist, René Pischel and Andrey Maksimov.

Finally, the authors wish to express their heartfelt thanks to Isabelle Sourbes-Verger and Jaque Grinberg for reviewing the final draft of the report and providing invaluable suggestions and constructive comments. Of course, it remains that responsibility for errors and infelicities now rests with the authors.

Contents

1 Introduction 1
  1.1 Russia's Legacy in Space: Glorious Past, Unsettled Present 1
  1.2 Objectives and Structure of the Study 3
  References 4

2 The Internal Evolution of the Russian Space Programme 5
  2.1 Russia in Space: A Challenged Resurgence 5
    2.1.1 Exogenous Factors............................... 7
    2.1.2 Endogenous Factors.............................. 13
    2.1.3 Impact on Russian Space Activities/Highlighting
          the Consequences 19
  2.2 Evolution and Outlook 20
    2.2.1 The Institutional and Industrial Setting: Historical
          Evolution and Current Landscape.................... 21
    2.2.2 Space Policies and Funding 34
    2.2.3 Current Capabilities and Future Programmes........... 40
  2.3 Assessing Current Status and Future Prospects.............. 51
  References 52

3 The External Evolution of the Russian Space Programme 55
  3.1 Russia's International Posture: Drivers and Evolution 55
    3.1.1 Russia's Drivers for Space Cooperation 55
    3.1.2 Types of Space Cooperation....................... 57
    3.1.3 Russia's Evolving Attitude to Cooperation 60
  3.2 Evolution of Russia's Space Relations with Key Players........ 62
    3.2.1 Russia-U.S. Space Relations—Bound to Cooperate? 62
    3.2.2 Russia and China—Back-up Partners? 69
    3.2.3 Relations with India—Shrinking Rationales............ 75
    3.2.4 Relations with Other Actors and Groupings 81

3.3   Assessing the Trajectory of Russia's International Posture   89
References   90

4   Prospects for Europe   95
4.1   Europe-Russia Relations: The Background   95
        4.1.1   Political Relations: From the PCA to the Five-Principle
                    Russia Policy   95
        4.1.2   Economic Relations: Important Partners   99
        4.1.3   Issues and Prospects   101
4.2   Europe-Russia Space Cooperation: Business as Usual or
            Selective Engagement?   104
        4.2.1   Europe-Russia Space Cooperation Frameworks   104
        4.2.2   Sectoral Cooperation Overview   106
4.3   Evaluating the Impact of Russia's Evolution on Cooperation
            Prospects for Europe   114
        4.3.1   Consequences of Russia's Space Sector Evolution   114
        4.3.2   Consequences of Russia's International
                    Posture Evolution   117
        4.3.3   Possible Developments and Future Opportunities   119
References   128

5   Conclusions   131

Annex A: Russia's Space Firsts   135

Annex B: Legal and Policy Framework of Russian Space Activities   137

Annex C: Russia's Major Programmes Overview   141

Annex D: The Evolution of International Joint Ventures   145

Annex E: Past and Future Europe–Russia Space
                    Cooperation Timeline   149

Acronyms

AA	Association agreement
ATV	Automated transfer vehicle
BRICS	Brazil, Russia, India, China, South Africa
CD	Conference on Disarmament
CIS	Commonwealth of Independent States
CNES	Centre National d'Etudes Spatiales (French Space Agency)
CNSA	China National Space Administration
COP21	21st Conference of the Parties
CSG	Guiana Space Centre
CTC	Cosmonauts Training Centre
DRDO	Indian Defence Research Development Organisation (India)
DSG	Deep Space Gateway (U.S.)
E3P	European Exploration Envelope Programme
EARS	EUMETSAT Advanced Retransmission Service
EC	European Commission
EDM	Entry, Descent and landing demonstrator Module
EEAS	European Union External Action
EELV	Expendable Launch Vehicle programme (U.S.)
EEU	Eurasian Economic Union
ELN	European Leadership Network
EO	Earth observation
ERA	European Robotic Arm
ESA	European Space Agency
ESPI	European Space Policy Institute
EU	European Union
EUMETSAT	European Organisation for the Exploitation of Meteorological Satellites
EVA	Extra-vehicular activity
FASO	Federal Agency for Scientific Organisations
FGUPs	Federal State Unitary Enterprises

FSA	Federal Space Agency
FSP	Federal Space Programme
FSP-2025	Federal Space Programme of the Russian Federation for 2016–2025
FTAs	Free-trade areas
FTP	Federal target programme
GCOS	Global Climate Observing System
GDP	Gross domestic product
GEO	Geostationary orbit
GLONASS	Global navigation satellite system
GLONASS-2020	Federal target programme on the development, utilisation and maintenance of the GLONASS for 2012–2020
GPS	Global positioning system
GSC	Guiana Space Centre
GSLV	Geosynchronous Satellite Launch Vehicle (India)
IAC	International Astronautical Congress
IBMP	Institute for Bio-Medical Problems of RAS
ICBMs	Intercontinental ballistic missiles
ICT	Information and communication technology
ILS	International Launch Services
IRNSS	Indian Regional Navigation Satellite System
ISA	Iranian Space Agency
ISC	International Space Company
ISON	International Scientific Optical Network
ISRO	Indian Space Research Organisation
ISS	International Space Station
JAXA	Japan Aerospace Exploration Agency
JSC	Joint-stock industrial companies
JUICE	JUpiter ICy moons Explorer
JV	Joint venture
KARI	Korea Aerospace Research Institute
KhSC	Khrunichev State Space Research and Production Space Centre
KSLV	Korean Space Launch Vehicle
LEO	Low Earth orbit
LMSM	Lunar Mission Support Module
LOP-G	Lunar Orbital Platform-Gateway (NASA)
LPSR	Lunar Polar Sample Return
LPSR	Lunar Polar Sample Return Mission
LRO	Lunar Reconnaissance Orbiter of NASA
MEK	Multi-element interplanetary expeditionary complex
MEO	Medium Earth orbit
MGNS	Mercury Gamma-ray and Neutron Spectrometer
MLM	Multipurpose laboratory module
MoU	Memorandum of understanding

MSL	Mars Science Laboratory of NASA
MTCR	Missile Technology Control Regime
NASA	National Aeronautics and Space Administration
NATO	North Atlantic Treaty Organisation
NII KP	Research and production establishment for the COSPAR equipment
NII TP	Research Institute of Electronic Instruments
NOAA	National Oceanic and Atmospheric Administration
NPO	Research and production establishment
NPO Energomash	Glushko Energomash research and production establishment
NPO IT	Research and production establishment of measuring equipment
NPO PM	Reshetnev applied mechanics research and production establishment
OECD	Organisation for Economic Cooperation and Development
OKB MEI	Special Design Bureau of the Moscow Power Engineering Institute
PCA	Partnership and cooperation agreement
PILOT	Precise Intelligent Landing using On-board Technology (ESA)
PNT	Positioning, navigation and timing
PROSPECT	Platform for Resources Observations and in Situ Prospecting in Support of Exploration, Commercial exploitation and Transportation (ESA)
PTK Federation	Piloted Transportation Spacecraft "Federation"
R&D	Research and development
RAS	Russian Academy of Sciences
RKK Energia	S. P. Korolev Rocket and Space Corporation Energia
RNII KP	Russian Research Institute for Space Instrument Engineering
ROSHYDROMED	Federal Service for Hydrometeorology and Environmental Monitoring
RSC	Roscosmos State Corporation
SB	Sub-programme
SC	State corporation
SCO	Shanghai Cooperation Organisation
SIS	Space International Services
SLS	Space Launch System (U.S.)
SPECTRUM	Space Exploration Communication Technology for Robustness and Usability between Missions (ESA)
SPM-1	Science and Power Module-1
SRI	Space Research Institute of RAS
STEM	Science, technology, engineering and mathematics
TAS	Thales Alenia Space
TGO	Trace gas orbiter
TsENKI	Centre for Operation of Space Ground-Based Infrastructure

ULA	United Launch Alliance
UM	Node module
UN	United Nations
UNEP	United Nations Environmental Programme
URSC	United Rocket and Space Corporation
USAF	United States Air Force
USSR	Union of Soviet Socialist Republics
VKS	Military space command
WMO	World Meteorological Organization
WTO	World Trade Organization

List of Figures

Fig. 2.1   Orbital launch event 2000–2017 (*source* ESPI Database)   6
Fig. 2.2   Global satellite industry output in mass
(*source* ESPI Database)   7
Fig. 2.3   Russia's GDP Evolution (*source* The World Bank, 2017)   8
Fig. 2.4   Depreciation of the Russian rouble (*source* Bloomberg,
2017)   9
Fig. 2.5   Russia's civil space budget   10
Fig. 2.6   Putin's approval rate (*source* Levada Center, 2017)   12
Fig. 2.7   Russia space sector crisis: causes and consequences   20
Fig. 2.8   Russian space agency timeline   25
Fig. 2.9   The evolution of Russia's space workforce   27
Fig. 2.10   Human resource breakdown (by typology and age Cohort)   28
Fig. 2.11   Structure of the Russian academy of sciences   29
Fig. 2.12   Space policy-making in Russia   35
Fig. 2.13   Russia's federal space programmes   36
Fig. 2.14   FSP-2025 annual budget composition (in Billion RUB)   38
Fig. 2.15   Budget breakdown of the FSP 2025   39
Fig. 3.1   Evolution of US-Russia cooperation on RD-180 engines   67
Fig. 3.2   Russia's regional space cooperation in Asia (*credit*
Volynskaya, 2014)   84
Fig. 4.1   President Putin and EU high representative Federica
Mogherini. *Credit* RaiNews 24   98
Fig. 4.2   The EU's five-principle policy towards Russia   98
Fig. 4.3   Main trade partners 2015. *Source* European Commission
(2017b)   100
Fig. 4.4   EU-Russia trade composition. *Source* European
Commission (2017b)   101
Fig. 4.5   Survey on Europe's sanctions. *Source* European
Leadership Network (2016)   102
Fig. 4.6   Survey on possible EU's posture vis-a-vis Russia.
Source European Leadership Network (2016)   103

Fig. 4.7   Soyuz Launch from GSC. *Credit* Sputnik News. 109
Fig. 4.8   Artist's view of ExoMars. *Credit* ESA 112
Fig. 4.9   ExoMars 2020 mission elements. 113
Fig. 4.10   The European robotic arm. *Credit* ESA 125

List of Tables

Table 2.1   Russia's launch failures 2010–2017 (*source
              ESPI Database*)   6
Table 2.2   Federal space programme budget 2017–2019
              (in billion RUB)..................................   10
Table 2.3   Heads of the Russian Space Agency   19
Table 2.4   Russia's new space ecosystem......................   33
Table 2.5   FSP-2025 budget as approved in March 2016   37
Table 2.6   Russia's current launch vehicles......................   41
Table 2.7   Launch vehicles under development.....................   42
Table 2.8   Russia's launch infrastructure.........................   43
Table 2.9   Development phases of Vostochny Cosmodrome...........   44
Table 2.10   Roscosmos earth observation missions in FSP-2025   45
Table 2.11   Roscosmos communications satellites as of 2016..........   46
Table 2.12   Roscosmos telecommunication satellites in FSP-2025   46
Table 2.13   Future GLONASS system   46
Table 2.14   Planned ISS modules and DSG work...................   48
Table 2.15   FSP-2025 science missions   50
Table 3.1   Drivers of Russia's space cooperation   57
Table 3.2   Relevant technology transfers from Russia in the field
              of launch vehicles   58
Table 3.3   Russia's international joint ventures for launch services......   59
Table 3.4   Total cost of Soyuz seat per year......................   63
Table 4.1   EU-Russia common spaces   96
Table 4.2   Trade in goods 2014–2016, €billions...................   100
Table 4.3   Trade in services 2013–2015, €billions   100
Table 4.4   Foreign direct investment, €billions   101
Table 4.5   The list of the launches from CSG including planned
              missions...   110

Table 4.6   ESA-Roscosmos ExoMars cooperation 111
Table 4.7   ESA-Roscosmos future lunar missions. 122
Table 4.8   Mapping of ESA contributions to Russian missions. 123
Table 4.9   Future ESA-Roscosmos space sciences and exploration
             missions. 123

Chapter 1
Introduction

1.1 Russia's Legacy in Space: Glorious Past, Unsettled Present

Over 60 years ago, the Soviet Union brought humanity into the space age with the success of Sputnik-1. Impelled by the logic of the Cold War and their technologically-founded vision for the future of mankind, Soviet leaders invested formidable human and technological resources to develop unequalled capabilities and unique expertise in what was deemed the final frontier of global geopolitics. Even though the one-upmanship contest to the Moon eventually underscored the victory of their American rival, from the mid-1950s to the early 1990s the Soviet space programme de facto remained an undisputed metric for worldwide space developments, as evidenced by its long list of pioneering achievements, covering the whole spectrum of space activities: the first probes to the Moon, Venus and Mars, the first EVA in 1965, the first rendezvous and docking between two manned vehicles in Earth orbit and exchange of crews, the first orbital laboratory, Salyut-1 in 1971, and the first permanently inhabited station, Mir, which orbited the Earth for 15 years, just to name the most remarkable (see Annex A for a more detailed overview).

All changed with the collapse of the Soviet empire in 1991. Like other industrial sectors of the Soviet Union, space faced a severe crisis, and was arguably the most impacted sector because of the dramatic reduction in political and ideological support, an essential factor having fuelled the "golden age" of Soviet space activities. In addition, the redefinition of the state borders of the former USSR led to the dispersion of several key elements of Russian space infrastructure and capabilities across independent sovereign states, particularly in Ukraine and Kazakhstan. In urgent need of hard currency to keep the programme running, and of an adequate activity level to maintain its industrial base, the Russian Federation opened rapidly and widely to foreign partners, greatly interested in buying its hardware and expertise, especially in the field of space transportation (both manned and unmanned). Considering that, at the time, there was only a handful of space-faring nations, and that the products of the former Soviet Union combined the advantages of demonstrated reliability and lower costs, Russia fortified *buyer-seller* relationships; not only with emerging spacefaring nations such as China and India, but also with established European and American

© Springer International Publishing AG, part of Springer Nature 2019
M. Aliberti and K. Lisitsyna, *Russia's Posture in Space*, Studies in Space Policy 18,
https://doi.org/10.1007/978-3-319-90554-9_1

players, both determined to benefit from Russia's cost-effective know-how, safeguard their interests in commercial markets, and prevent possible proliferation of critical technologies to "rogue states". During the 1990s, Russian and Western partners established several joint ventures to commercialise the former Soviet launchers and converted ICBMs. As documented in a previous ESPI report, "those joint ventures enabled the sector to maintain a level of activity necessary for its survival. At the same time, the lower priority given to space and limited funding for more than a decade led to the decay of the Russian space systems, in particular civil satellite systems. As a result, in 2000 the GLONASS navigation constellation had only eight of the nominal 24 satellites, and few years later, Russia no longer had a single meteorological satellite"(Mathieu, 2008).

After this decade-long crisis, things changed again with the turn of the new millennium. The new political leadership guided by President Vladimir Putin gave new impetus to the development of the country's space activities and put the sector back among the top priorities of Moscow's domestic and foreign policy agenda. Supported by the progressive recovery of Russia's economy, renewed political stability and Russia's improving external environment, at the beginning of the century, the programme started to re-assert strong ambitions and the resolve to regain its original position on the international scene (Harvey, 2007). Towards this, several major space programmes were adopted, including the Federal Space Programme 2006–2015 (comprising new scientific and space exploration missions, EO and telecommunications systems, and the ISS deployment); the Federal Target Programme on the development of Russian cosmodromes 2006–2015; and the Federal Target Programme on the redeployment of GLONASS 2002–2011. These three ambitious programmes received further stimulus during President Putin's second mandate (2004–2008), when he made the space sector one of the main priorities for the development of Russia's industry, together with aeronautics and ship-building (Mathieu, 2008). This renewed commitment to the development of space activities was duly reflected in the sharp increase in the country's space budget throughout the decade.[1] Thanks to the funds made available by flourishing energy exports, Russia's space expenditure continued to grow even in the midst of the global financial crisis. Suffice it to say that between 2008 and 2009 the civil space budget alone doubled in a single year to reach Rub 88.64 billion (approximately €2billion) (European Space Agency, 2010). Besides new programmes and increased funding, the spectrum of activities was also widened to encompass a new focus on space applications and commercial products. Moreover, in order to increase competitiveness and the valorisation of the benefits derived from the country's capacities and expertise, the space industry underwent an ambitious restructuring, which combined stronger state involvement with the desire to create national champions in each major branch (propulsion, launchers, space systems, etc.).

---

[1]Given the rather modest budget allocated throughout the 1990s, this increase should not be overstated either. Russia's space budget remained—and still is—much lower than that of the major spacefaring nations, including the European Space Agency (ESA).

All in all, whereas the ambitious goals setting forth the future development of the Russian space programme were supported at the highest political level, ten years later—for the 60th anniversary of the launch of Sputnik—much of this prospected resurge seems to have failed the reality check with the ambitious targets set by the political leadership.

Negatively impacted by many recent endogenous and exogenous developments, such as the excess in manufacturing capacity and low labour productivity as well as the progressive worsening of its economic performance and external relations with Western countries, Russia's posture in space has had to be once again adapted to cope with new realities, causing some major reconsiderations on both the domestic and international fronts. These impacts are well reflected in the readjusting of Russia's space funding and programmes, in the 2015 reorganisation of Roscosmos, as well as in the changing attitudes towards international space partnerships. One of the most striking features in this respect is that Russia, which had increasingly opened its space programme to Europe and the United States since the end of the Soviet Union, has in the past few years shown greater resolve to achieving autonomy from foreign sources, while also enlarging its international outreach to new emerging spacefaring nations and groupings such as the BRICS.

Needless to say, these evolutions and changing attitudes have the potential to strongly impact current and upcoming relations with foreign partners, including Europe, which has traditionally considered Russia as one of its two main strategic partners in its space endeavour. Whereas long-standing cooperation has hitherto developed between Russian and European partners in various areas, from science to launchers to human spaceflight, the ongoing redefinition of priorities and directions raises a number of open questions for Europe.

1.2 Objectives and Structure of the Study

Drawing on extensive research, this study will shed light on the internal and external evolutions of the Russian space programme and sector, with the goal of providing an assessment of their impact on current and foreseeable Europe-Russia space cooperation. The underlying research questions driving the reflections of this study include:

- What are the most relevant evolutions the Russian space sector has undergone over the past 10 years?
- What are the major changes in its international posture, and what factors have been triggering such changes?
- How are these internal and external evolutions in Russia's attitude impacting Europe-Russia (space) cooperation?
- How strong is the need and will for Europe to continue cooperating with Russia, and in which fields?

The book is comprised of three major chapters. In the first part, the study will assess the domestic evolution of the Russian space programme over the past ten years with the goal of explaining its current status and future outlook. The chapter is organised in three major sections. In the first part, the factors that have shaped Russia's recent posture in the space industry are identified and explained. In the second part, the chapter looks at the evolution of the programme itself and its status in 2017. Special attention is given to the organisational setting, space policies and policies, and the programme. Finally, an assessment of the status and outlook of Russian space activities is provided.

In the second part, the book investigates the evolution and current state of affairs Russia's international space posture, with a focus on its attitude towards international cooperation with key partners. This analysis will be carried out in relation to the evolution of both domestic and international dynamics that have been impacting the country's posture in space over the past few years. Accordingly, the chapter will first provide a concise overview of the evolution of Russia's space diplomacy, and then review past and current cooperation undertakings by analysing the rationales, drivers, limitations and possible evolution of such relations, from cooperation to competition (and vice versa), in different space-related areas. In this part, perspectives on Russia's international space relations will be intertwined with broader political considerations.

In the third part, the book will examine the possible consequences of these domestic and international developments for Europe. The final chapter will more specifically assess the current Europe-Russia interplay in both the political and space arenas, and subsequently evaluate the potential impact of recent changes on Europe's cooperation with its Russian partner. Finally, different options for cooperation between Europe and Russia will be defined and assessed to be able to maximize Europe's opportunities, while minimising risks.

References

European Space Agency. (2010). *European space technology master plan*. Noordwijk: European Space Agency.

Harvey, B. (2007). The rebirth of the Russian space program. 50 years after Sputnik, New Frontiers. Chichester: Springer—Praxis.

Mathieu, C. (2008). *Assessing Russia's space cooperation with China and India: Opportunities and challenges for Europe*. Vienna: European Space Policy Institute.

Chapter 2
The Internal Evolution of the Russian Space Programme

2.1 Russia in Space: A Challenged Resurgence

In the mid-2000s, there were many expectations about the prospects of the Russian space programme. For the majority of space policy analysts, the decade-long crisis of the 1990s was over and the rebirth of the Russian space activities appeared imminent. After all, reasons for optimism were manifold and included:

- A restored funding capacity and a growing budget;
- A resurging annual launch rate;
- A growing presence in the global commercial market (particularly in the launcher domain);
- An expanding level of ambition with approval of several major programmes;
- A central role for the programmes of many established and emerging spacefaring nations, including the United States;
- Strong support at the highest political level.

In 2007, President Vladimir Putin openly committed to making the space sector a strategic priority for Russia. Ten years later, on the 60th anniversary of the launch of Sputnik, the country's space programme has not experienced the anticipated resurgence and actually seems to wallow in a state of systemic crisis. Together with the delay—if not cancellation—of many flagship programmes, the most visible sign of this crisis is perhaps the embarrassing string of 18 launch failures that have taken place over the past six years, which have amounted to billions of roubles in losses (see Table 2.1).

Other important indicators of Russia's space crisis are the reduction in the share of worldwide launch activity over the past 10 years (see Fig. 2.1) as well as the reduction of the output of the Russian satellite industry, which is even more striking when excluding the human spacecraft Soyuz and Progress—since they correspond to a substantial share of Russian industrial output (see Fig. 2.2).

A plethora of challenging factors can be identified to explain the failed resurgence of the Russian space programme. These can be grouped into exogenous and endogenous factors.

© Springer International Publishing AG, part of Springer Nature 2019
M. Aliberti and K. Lisitsyna, *Russia's Posture in Space*, Studies in Space Policy 18,
https://doi.org/10.1007/978-3-319-90554-9_2

Table 2.1 Russia's launch failures 2010–2017 (*source* ESPI Database)

Date	Launcher/stage	Payloads	Notes
5 December 2010	Proton M/Block DM03	GLONASS-M (x3)	Wrong orbit
1 February 2011	Rokot/Briz-KM	GEO-IK-2	Wrong orbit
18 August 2011	Proton M/Briz-M	Ekspress AM4	Wrong orbit
24 August 2011	Soyuz-U	Progress M-12M	Failed to reach orbit
23 December 2011	Soyuz-2.1b/Fregat	Meridian No. 5	Failed to reach orbit
6 August 2012	Proton/Briz-M	Telkom-3 Ekspress-MD2	Wrong orbit
8 December 2012	Proton-M/Briz-M	Yamal-402	Partial failure
15 January 2013	Proton M/Briz-M	Cosmos (x3)	Partial failure
31 January 2013	Zenit	Intelsat-27	Wrong orbit
2 July 2013	Proton-M/DM-03	GLONASS-M (x3)	Failed to reach orbit
16 May 2014	Proton/Briz-M	Ekspress-AM4R	Failed to reach orbit
27 April 2015	Soyuz-2.1a	Progress M-27M	Failed to reach orbit
16 May 2015	Proton-M/Briz-M	MexSat-1	Failed to reach orbit
5 December 2015	Soyuz-2.1v/Volga	Kanopus-ST	Partial failure
1 December 2016	Soyuz-U	Progress MS-04	Failed to reach orbit
14 July 2017	Soyuz-2.1b/Fregat	Various cubesats (x9)	Partial failure
28 November 2017	Soyuz-2.1b/Fregat	Meteor-M 2-1	Failed to reach orbit

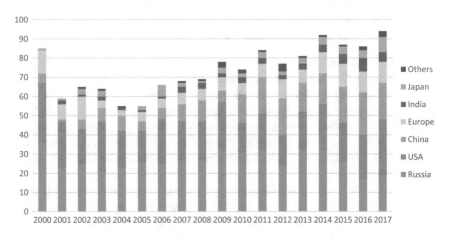

Fig. 2.1 Orbital launch event 2000–2017 (*source* ESPI Database)

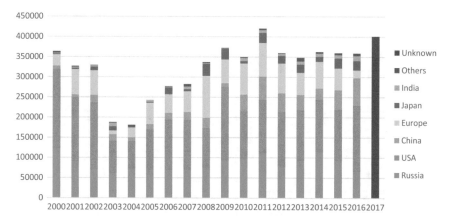

Fig. 2.2 Global satellite industry output in mass (*source* ESPI Database)

2.1.1 Exogenous Factors

As with all spacefaring countries, Russian space activities cannot be assessed in isolation from the broader socio-economic and political context in which they are embedded. Indeed, the primary factors that have thwarted a second golden age in Russian space activities are to be found in a series of exogenous factors outside the direct control of the country's space authorities. These exogenous issues are mainly related to a set of drawbacks in Russia's broader political and economic environment, including:

- An unstable political environment (conflict in Ukraine, military annexation of Crimea, Syria, Iran)
- Escalation of the Ukrainian conflict
- Tense political relations with Western countries over key issue-areas (Ukraine, Syria, Iran, North Korea)
- A financial crisis triggered by the collapse of the Russian rouble in 2014
- Negative economic performance resulting from the oil crisis and Western economic sanctions.

As explained below, the impact of these factors has not been limited to budgetary aspects, but has also extended to programmatic ones as well as commercial activities and cooperative undertakings.

2.1.1.1 Economic Factors

Starting with economy-related issues, Russia entered the second decade of the 21st century in the midst of the worldwide economic crisis. In 2009, the Russian economy shrank by 8.5%, amid declining oil revenues and the departure of foreign capital from the country. After a swift recovery in 2010–2013, Russia entered an even more

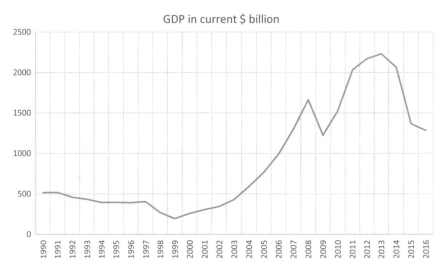

Fig. 2.3 Russia's GDP Evolution (*source* The World Bank, 2017)

menacing financial crisis that caused a severe contraction of its GDP (see Fig. 2.3) (The World Bank, 2017). In the span of a couple of years, Russian GPD has contracted by roughly $300 billion and has returned to approximately the same level as in 2007, though it is slowly recovering now.

To a large extent, Russia's still ongoing financial crisis is the direct result of the collapse of the Russian rouble and associated speculations that began in the second half of 2014. A decline in confidence in the Russian economy caused investors to sell off their Russian assets, which in turn led to a decline in the value of the Russian rouble and sparked fears of a Russian financial crisis (see Fig. 2.4) (Bloomberg, 2017).

Investors' loss of confidence in the Russian economy stemmed from at least two major sources. The first was the fall in the price of crude oil, which is a major export of Russia (following its yearly high in June 2014, crude oil has declined in price by nearly 50%). The second was the result of international economic sanctions imposed on Russia following Russia's annexation of Crimea and the Russian military intervention in Ukraine (see further).

Inevitably, these unfavourable economic conditions have heavily impacted the space sector, thus further exacerbating its perceived state of crisis:

- Russia's current economic problems have led to a reduction of the budgetary allocations for the country's space activities of the order of 23%, thereby affecting several major programmes.
- Despite the depreciation of the rouble, Russia's economic problems have harmed the country's exports on commercial markets (and more broadly all operations involving foreign partners).

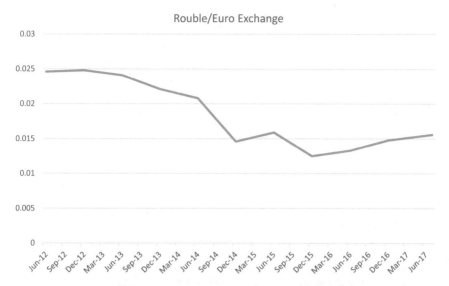

Fig. 2.4 Depreciation of the Russian rouble (*source* Bloomberg, 2017)

Whereas in the years up to 2014 Russia's space budget experienced steady growth, reaching a peak of 128 billion roubles (€5 billion),[1] with the recent economic downturn space spending has begun to undergo repeated assaults by the Ministry of Finance during budgetary planning. Despite several complaint letters from Roscosmos to Russian President Vladimir Putin, overall space programme funding was subject to a 10% reduction per annum in the period 2014–2016 (see Fig. 2.5).

The impact actually went beyond a short-term budget reduction, and affected the new 10-year master plan for Roscosmos. In 2014, when efforts to prepare the Federal Space Programme 2016–2025 began, the Russian government initially pledged a 2849 billion roubles budget (€63.31 billion) over the period. The proposal, however, underwent a number of revisions from various government institutions before its approval.[2] As a consequence of the budget cuts requested by the Ministry of Finance

---

[1]Thanks to government subsidies, the Russian space industry weathered the 2008–2012 economic crisis relatively unscathed. During 2012–2015, the Russian government promised to invest 650 billion rubles into the space industry. (According to data released by the Ministry of Economic Development in September 2012, a total of 590 billion rubles was promised for the Russian space program during 2013–2015).

By the end of 2012, the Russian government promised to spend 2.1 trillion rubles (including non-federal funds) before 2020. In April 2013, President Putin quoted 1.6 trillion rubles ($51.8 billion) to be spent on the space program until 2020. The revival of the GLONASS navigation network and the construction of the new launch site Vostochny were listed as the biggest space budget items.

[2]Throughout 2015, Roscosmos worked to push the 2016–2025 FSP through around 20 ministries and institutions, including the Russian Academy of Sciences, the Ministry of Defence, the Ministry of Finance and various users of the programme, before its final approval by December of that year. In addition, Roscosmos also worked to approve its annual budget and projected funding for two

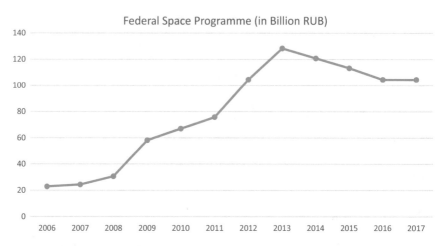

Fig. 2.5 Russia's civil space budget

Table 2.2 Federal space programme budget 2017–2019 (in billion RUB)

	Approved March 2016	Revised December 2016	Variation in Billion RUB	Variation in percentage (%)
2017	104.5	92.5	−12	−11.6
2018	104.5	89.2	−14.8	−14.7
2019	104.5	86.3	−18.2	−17.6
Total	313.5	268	−45	−14.6

to face the challenging economic situation, in April 2015, the initially proposed budget was reduced by 800 billion roubles (−28%; €14 billion), thus reaching the level of 2004 billion roubles (€37.81) over 10 years. And in the final version, approved in March 2016, it stood at a meagre 1406 billion roubles (−50%; €18.5 billion), with an average annual earmarked budget of 104 billion roubles (€1.8 billion as of March 2016).

These figures were further revised in December 2016. When approving the three-year budget period covering 2017, 2018 and 2019, the government eventually cut an additional 45 billion roubles ($1.02 billion) from Roscosmos' budget (see Table 2.2). Although both the Ministry of Finance and President Putin himself have pledged an extra 115 billion roubles (€ 1.6 billion as of December 2017) for the period 2022–2025, some critics have noted that this reduction will now create a substantial gap between the official cost of formally approved projects and the actual budget available for Roscosmos within the FSP-2025. The actual impact of budget reductions for the years 2017–2019, however, has not been officially disclosed by Roscosmos.

---

subsequent years. By the end of 2015, the 2016 budget was slated to see an increase in comparison to the funding for 2015, however that increase would not be as high as first projected in the three-year forecast that had been made in 2014.

It is also worth noting that the budget foreseen for the FSP-2025 has actually doubled in absolute terms compared to the previous 10-year master plan. However, when considering the exchange rate with Western currencies (US$ or €), the current budget of €1.7 billion per year has substantially decreased in comparison to the 2012/2013 period, when it peaked at the level of roughly €5 billion. Thus, even though domestic space activities might not be highly affected, any operation involving transactions with foreign entities can be expected to have a burdensome impact on Russian sheets.

Besides this impact on budget and programmes, the devaluation of the Russian currency has also generated a seemingly paradoxical consequence for the country's commercialisation efforts on global space markets (including, in particular, for launchers). Indeed, while in principle the depreciation of the rouble should have improved the competitiveness of Russia's commercial solutions, this has not been the case because, as noted in Sect. 2.1.1, Russia is today highly dependent on imports, particularly electronic components (roughly 80% of imported components). As the currency weakens, costs continue to rise, with the space sector also losing its competitive advantage in terms of low labour costs due to high inflation rates. Therefore, despite the currency devaluation, launch vehicles (like other systems offerings) can no longer be offered at very low prices.

2.1.1.2 Political Factors

In addition to the country's economic problems, Russia's space activities have been suffering from a number of strains in the political sphere. To Russia's credit, it must first be recognised that the country has benefited from internal political stability since the first election of President Vladimir Putin in 2000. In May 2008, after two presidential mandates, President Putin had to hand over the presidency to Dmitry A. Medvedev. He became Prime Minister and leader of "United Russia", Russia's major political party, actually reinforcing his position and influence in the Duma and setting the stage for his return as President in 2012.

By concentrating power within the Kremlin,[3] President Putin has been able to secure a high degree of internal political stability and, despite the recent economic downturn, his popularity has remained highly stable (see Fig. 2.6) thus most likely securing his reconfirmation at the 2018 elections (Levada Center, 2017).

However, on the international scene, Russia has been confronted with an increasingly difficult situation. Following the 2008 war in Georgia and the ever-growing tensions with the members of the North Atlantic Treaty Organisation (NATO) over the missile defence system to be installed in Poland and the Czech Republic, Russian relations with Western countries have remained characterised by a high degree of mistrust and political confrontation.

---

[3]Although this concentration of power has made Western observers concerned about the risks of authoritarianism, as explained by several comparative scholars, it should not be interpreted in light of authoritarianism, but rather as an effort to increase state-ness (intended as both capacity and autonomy), following the de-statification process that ensued after the fall of the Soviet regime.

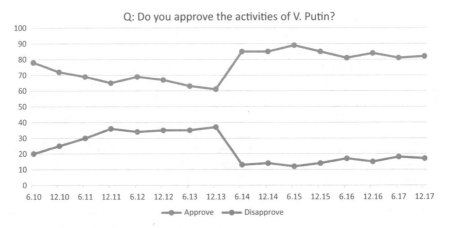

Fig. 2.6 Putin's approval rate (*source* Levada Center, 2017)

Despite the Obama Administration's attempts to "press the Reset Button" in 2010, U.S.-Russia political relations steadily worsened, owing to a number of diplomatic irritants as well as Washington's perceived attempts to reduce Russia's international role to that of a regional, rather than global, power. The European Union's diplomacy too, with its efforts to foster closer economic and political ties with ex-Soviet countries, was seen as an intrusion into Russia's sphere of interest. Not surprisingly, the proposed Ukraine-EU Association Agreement eventually turned into a source of strategic rivalry that almost stretched to open conflict.

The Russian military intervention in Ukraine, which began in late February 2014 and culminated in the annexation of the Crimean Peninsula, prompted a number of governments and international organisations to apply sanctions against Russian individuals, businesses and officials. Sanctions were approved by the EU, the U.S. and other countries, including Canada, Japan and Australia, contributing to Russia's financial crisis.

While the new context did not directly impact the course of Russia's civil cooperation with Western countries (i.e. ISS operations and bilateral cooperation with the European Space Agency—ESA),[4] it nonetheless generated other important consequences.

For one thing, all the military and civil cooperation between Ukrainian entities and Russia was immediately suspended by the newly elected president Petr Poroshenko, inter alia causing serious drawbacks for the operations of Russo-Ukrainian joint ventures such as the International Space Company (ISC) Kosmotras and Space International Services (SIS) Ltd (see Chap. 3 for a more detailed assessment).

In addition, the combined effect of the Crimea crisis and wide economic sanctions led to an increasingly isolated position of Russia and contributed to triggering the

---

[4]See Chaps. 3 and 4 for a more detailed assessment on the impact of the new political context.

above-discussed financial crisis, thus indirectly undermining the budget allocations for the space programme.

Furthermore, in response to Russian actions in Ukraine, the EU and the U.S. enacted export restrictions on many sensitive technologies such as electronics, which—as discussed—were of crucial importance for the conduct of Russian space activities. Consequently, Russia had to begin a costly effort to replace foreign components in its space programme, hence further aggravating the already shrinking spending possibilities. Roscosmos estimated that more than a billion roubles (then €700 million) would be necessary over the 2014–2018 period to free Russia from dependence on Ukrainian and Western raw materials/components and establish domestic assembly and production. Considering that for some of these components no domestic technology was available, in 2014 Russia decided to put in place an import substitution programme to increase technological non-dependence. In addition, to counter the immediate shortfalls, Russia also started to purchase electronic components from China (with a deal estimated worth several billion US$ in early 2015).

More broadly, it must be noted that recent strains in the political sphere have contributed to altering Russia's posture in the international space arena and its attitudes towards space cooperation with key countries. These changes will be analysed in greater detail in Chap. 3.

2.1.2 Endogenous Factors

Apart from the exogenous issues discussed above, the resurgence of the Russian space programme has been thwarted by several structural factors in its space industry. Despite the positive rhetoric from the government, during the expansion phase of the 2000s the Russian space sector proved unable to tackle the many structural problems inherited from the Soviet period or those resulting from the difficult socio-economic situation faced by Russia during the 1990s. Remarkably, these deep-rooted problems were not only manifold, but also "nested" in nature, meaning that they were closely interlinked with each other. In a nutshell, the most prominent issues identified through a review of publicly available documents and interviews with stakeholders were related to:

- Gaps in quality control/lack of accountability
- Weak satellite components base/lack of production material
- Out-dated production base
- Inefficient use of budgetary funds
- Scattered industrial capacity/spiralling overhead costs
- Misuse of funds/endemic corruption
- Excess in manufacturing capacity/low labour productivity
- Brain drain
- Lack of innovation
- Multiple reorganisations in the orgware and leadership
- Lack of vision.

Although the substantial budget increase that occurred throughout the 2000s enabled these problems to be circumvented, or at least their most visible effect silenced, they nonetheless, persisted and eventually started to dramatically resurface in the aftermath of the above-mentioned string of launch and spacecraft failures.

2.1.2.1 Lapses in Quality Control

The first set of problems that came to the fore during the investigation process into the launch failures that started in 2010 was related to the lack of oversight and inherent difficulties in maintaining quality control standards (Toporov, 2017).[5] In addition, the investigation process brought to the light the problems related to the lack of accountability by industrial players. Indeed, the companies responsible for flawed parts and equipment had no financial or other types of responsibility, with the government bearing all costs related to failures. Whereas this is understandable considering that these companies were state-owned, it was also clear that the lack of serious consequences had not provided incentives to avoid the shortcomings responsible for launch failures. These issues triggered the immediate initiation of reform actions to ensure improved quality monitoring. However, as the launch failures continued in subsequent years despite increased quality controls at the industrial level, it became evident that the problems confronting Russian space industries were "structural" in nature and that these problems could no longer be ignored.

2.1.2.2 Weak Components Base

Among the identified structural problems, the most visible were related to a lack of raw material production and an obsolete subsystems and components' base for the manufacturing of space systems. Notwithstanding that the resolution approving the Federal Space Programme 2006–2015 had put special emphasis on reducing dependence on foreign components, over the previous 10 years the country's space industry had continued to lose its capability to produce many satellite components. As reported by ESPI in a previous study, "in 2010 the inferior quality of electronic components sourced by the Russian space industry was addressed at the collegiate council of the Russian space agency, Roscosmos. Continued reliance on those components increased the number of failures both at the AIT stage and in the operational use of satellites and other space technology. The problem seemed to be a lack of ground testing and too much reliance on calculations instead of full-scale testing" (Al-Ekabi, 2016). Not surprisingly it was estimated that by the end of 2013 up to 80% of the equipment on new Russian satellites had been imported from foreign sources (in particular from Thales Alenia Space and MDA) (Reuters, 2013).

---

[5]The situation could be partially explained by the organisational changes in the quality control processes put in place during the 2000s.

2.1.2.3 Out-Dated and Scattered Space Infrastructure

Another more systemic cause seemed associated with the increased obsolescence of the industrial infrastructure, including, test beds, vacuum chambers, clean rooms, mechanical and electrical ground support equipment (EGSE/MGSE) of the Federal space agency Roscosmos (Zak, 2012). It was estimated that almost 90% of the industrial infrastructure in 2013 was older than 20 years. This problem was acknowledged in a 2013 report by the Russian Audit Chamber, which more broadly declared Russia's Federal Space Program as "ineffective", because of poor management of space activities and an inefficient utilisation of the funds allocated for space projects (RIANOVOSTI, 2013).

Much of this inefficiency was clearly linked to the scattered industrial landscape for space activities in Russia. Indeed, owing to the Soviet Union's goal to ensure the development of remote areas through the relocation of heavy industry from Western Russia to the Ural and Siberian regions, the Russian space industry remained characterised by a large number of state-owned large- and medium-size entities geographically spread all over the national territory. These horizontal structures prevalent under the Soviet regime then left the Russian government with the challenging task of overseeing a scattered industrial landscape, where even development and exploitation facilities, while typically co-located, were structured around separate entities, often with redundant tasks.[6]

2.1.2.4 Misappropriation of Funds and Corruption

This redundancy visibly contributed to generating spiralling overhead costs and, equally importantly, misappropriation of funds or corruption problems. Corruption is considered widespread in Russia's economy, and the space sector is no exception. There are several cases of managers and workers arrested for bribery as well as instances in which a substantial part of the governmental space budget has been diverted from major programmes as a part of "overheads", thereby causing shortcomings, delays and spiralling costs.

A well-known example is the construction of the Vostochny cosmodrome, which has systematically undergone numerous public delays over the past 10 years owing to corruption issues, among others (The Moscow Times, 2017). As recounted by ESPI

---

[6]The Soviet space programme separated research, design and production institutions and industry based on a formal model called the 'research—production cycle', with successive steps (fundamental research, development and mass production) handled by separate specialised entities. The types of organisation can be placed in three broad categories: scientific research institutes (NIIs), design bureaus (KBs) and production facilities. In the 1970s, integrated networks were put in place to cope with military requirements imposed by the arms race by combining institutes, design bureaus and plants under Science and Production Associations (NPO). The collapse of the Soviet Union led to the splitting up of this industrial organisation so that a joint stock company can now comprise an entire NPO or a single production plant, design bureau or research institute detached from its former network.

in its Space Policies, Issues and Trends report, in 2014, the cumulated delay required a cash infusion of an additional 50 billion roubles ($1.3 billion) to finish construction on time, but at the same time the head of a company involved in the construction of the spaceport was arrested and charged with embezzling 1.8 billion roubles ($43 million) from the project (Al-Ekabi, 2015). A more recent example is given by the January 2017 decision by Roscosmos to withdraw all second and third-stage engines for the Proton-M launch vehicle. Although the agency officially cited "technical reasons", some Russian media reported that factory managers manufacturing engines for Russia's Proton-M rocket may have been replacing precious metals with cheaper alternatives that were possibly the source of the failures of Proton' upper stage in the past few years (The Moscow Times, 2017).

2.1.2.5 Excess in Manufacturing Capacity and Low Productivity

Together with the corruption problem and the scattered industrial landscape, much of the inefficiency in the budget utilisation can be attributed to an excess in manufacturing capacity and low labour productivity.[7] Prior to the latest industrial consolidation (see Sect. 2.2.1), it was estimated that the capacity utilisation of the space sector infrastructure was less than 50%. As for the labour productivity, it was found to be roughly 2–4 times lower than in other spacefaring nations, highlighting a mismatch between the labour force and the activity level. In the early 2010s, 248,000 people were employed in the Russian space sector, 65% more than the optimal workforce of 150,000 estimated by Roscosmos. The optimisation policy pursed in recent years has given way to limited workforce reductions, at least in absolute terms. In fact, by 2015, the sector still counted 238,000 employees spread over 94 organisations. Yet, even in Russia, it is very difficult to optimise labour in state-owned organisations.

2.1.2.6 Brain Drain and Poor Labour Conditions/Poor Workforce Management

Notwithstanding the large number of employees in the sector, the country is facing huge strains in ensuring continuity of experience. To a large extent, this is because for a long time after the collapse of the USSR, Russia suffered from difficult economic conditions, which caused a purported inability to ensure employment stability and adequate remuneration, a subsequent loss of industrial know-how, and a shortage of skilled and young workforce. These problems led to the loss of important industrial capabilities, as in the case of the cancellation of the well-known Buran programme,[8] and more importantly, to a menacing brain drain. Indeed, as reported by many, "Rus-

---

[7] It has to be recalled that Russia's space industry was created for conducting more than 100 launches a year and no powerful computers were used back then.

[8] As highlighted by Alexey Beljakov, the head of the space technology cluster in Skolkovo, it would now be impossible to rebuild the famous Buran rocket, because the crucial knowledge of the

sia's space specialist population is aging, and their competence is waning due to the low attractiveness of space careers, with young people not entering—or even leaving—the industry to work abroad or in other sectors known for better conditions and salaries such as the ICT, oil and gas.[9]

2.1.2.7 Little Room for Innovation

A closely related set of problems in Russia's space sector is its decreasing innovation capacities. In the 2016 Space and Innovation Report, the OECD identified a set of conditions that are apposite for space innovation. The Paris-based organisation more specifically identifies: (a) an efficient system for knowledge creation and diffusion in institutions (from fundamental knowledge to technology transfer to other sectors); (b) policies that encourage innovation and entrepreneurial activity (i.e. grants, procurement mechanisms, prizes), while providing as much clarification as possible; (c) a business environment that encourages investment in technology and in knowledge-based capital (allowing experimentation with new ideas, technologies and business models); (d) a skilled workforce.

All these conditions are somehow wanting in the Russian space sector. As contended by several analysts, the Russian space industry in general has never been customer-oriented and motivated to produce competitive products and services. In the same vein, national space policy was oriented to supporting only major enterprises, giving them most R&D contracts. The structure of the Federal Space Programme, the major policy document defining the development of the national space sector, has historically been very rigid, making it hard to allocate funding for innovative research. Moreover, the structure was not friendly to technological transfer either. As a result, most breakthrough technologies, developed for the space programme (e.g. the Buran space shuttle) remained unused until they eventually became obsolete (Kosenkov, 2016).

As for space applications, which are one of the three overarching thrusts driving innovation in the sector,[10] national space assets have been typically usurped by the military and other governmental bodies. The dissemination of space-based data received from the Russian orbital constellation has faced multiple restrictive barriers and the absence of commercialization mechanisms, which have contributed to a loss of competitiveness of Russian satellites in terms of quality of provided data, lifetime, cost and manufacturing period (Kosenkov, 2016).

---

technological processes died with the old generation of engineers. More broadly, the problem of transferring knowledge remains very actual in Russia.

[9]This is due in part to reportedly low pay in the space sector. Also, some claim that, to comply with the Russian state secrets law, space workers are not allowed to travel outside of Russia—a big disincentive for young Russians.

[10]Together with the expansion of downstream space applications, the major thrusts for space innovation are the persistence of public investments to serve national security and scientific objectives, and the pursuit of human space exploration by both public and private actors.

The above-mentioned depletion of human resources and industrial technological competences has also undermined innovation capacities. In the early 2010s, the identified solution was to improve the institutional environment for space activities by making it more supportive of innovation through the establishment of public-private partnerships and open innovations frameworks. The overarching idea was that the creation of a competitive private space sector would substantially enhance the technology maturation rate and quality of provided products and services, thereby closing Russia's technological gap. Such reforms are still on-going but have yet to produce the desired outcome.

2.1.2.8   Multiple Reorganisations and Changes of Leadership

Another cause/symptom of the systemic crisis facing the Russian space sector can be found in the multiple reorganisations the orgware has gone through over the last decade.[11] A more detailed account of the transformations in the institutional setting will be provided in Sect. 2.1.1, but an important point to highlight is that, whereas these reforms were certainly intended to tackle the structural problems facing the Russian space sector (streamlining the industrial processes, eliminating excess manufacturing capacity and the inherent misuse of funds, improving innovation, etc.), it is undeniable that they have also contributed to creating uncertainty over responsibility and have impaired efficiency and stability, at least in the short term(see Sect. 2.1.1 for a more detailed analysis).

These multiple reorganisations have also been accompanied by changes in the leadership of important organisations, as a result of internal struggles. Most notably, over the past six years the heads of the Russian space agency have changed four times, leading to multiple changes of strategy and direction. Table 2.2 provides a list of the space agency's reorganisations and of its leadership since its creation in 1992 (Table 2.3).

2.1.2.9   Lack of Forward-Looking Vision

The multiple reorganisations and change of leadership are associated with another important cause of the systemic crisis of the Russian space sector: the lack of vision about the country's future role in space. To be sure, Russia's civil space activities are guided by long-term plans that define the programmatic priorities for the subsequent 10 years. The Federal Space Programme 2006–2015 was approved in 2006, whereas the Federal Space Programme 2016–2025 was approved in 2016. In addition, in April 2008, the then outgoing Russian president Vladimir Putin chaired a meeting of the Security Council (known in Russian as *Sovet Bezopasnosti*), with the goal of defining the programmatic goals of the Russian space programme until 2020.

---

[11]The term *orgware* refers to the organisational structures set up to develop and run technological *hardware*, in this case the space programme.

Table 2.3 Heads of the Russian Space Agency

	Director	Start	End
Russian Space Agency	Yuri Koptev	February 1992	May 1999
Russian Aerospace Agency	Yuri Koptev	June 1999	March 2004
Federal Space Agency	Anatoly Perminov	March 2004	April 2011
	Vladimir Popovkin	April 2011	October 2013
	Igor Komarov	January 2015	August 2015
	Alexander Ivanov	August 2015	December 2015
Roscosmos State Corporation	Igor Komarov	January 2016	Present

What is coming under question, however, is not the lack of planning, but the focus merely on the preservation of prior capacities, rather than the acquisition of new ones through forward looking programmes. As also highlighted by Mathew Bodner, "Russia's priorities in space are far more grounded than its Soviet predecessor. [...] The Soviet Union, an ideological superpower, had very clear reasons to push forward into space: Communism was humanity's future, they believed, and that future was in space. The Cold War gave additional ideological impetus, as space could demonstrate the superiority of their system. [...] Post-Soviet Russia is not an ideological nation. In many ways, it is a nostalgic nation. This nostalgia has been expertly co-opted by the government under President Vladimir Putin. Under him, Russians largely draw pride from looking back, rather than looking forward. And in this regard, the space program has already provided what it needs to" (Bodner, 60 years after Sputnik, Russia is lost in space, 2017).

Not surprisingly, the current Russian space programme, which is primarily focused on continuing Soviet-era programmes, proves to have lost the vision of its long-term directions. This discrepancy between the implementation of programmes and the lack of clear objective was best captured in April 2008 by Vitaly Lopota, then head of RKK Energia (chief developer of manned spacecraft). Speaking at a press conference, he told reporters that his organization could not formulate any concept of a new Russian spacecraft without knowing "where it would fly."

To a large extent, this loss of vision is closely associated with the current shortage of material means to purse new programmes, the cause of which has to be retraced to the adverse conditions outside the space sector.

2.1.3 Impact on Russian Space Activities/Highlighting the Consequences

As emerges from the previous sections, the combination of endogenous and exogenous factors has not only contributed to generating a state of system crisis in the

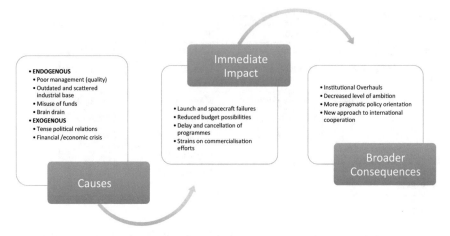

Fig. 2.7 Russia space sector crisis: causes and consequences

Russian space enterprise, but has also caused far reaching impacts on the country's space activities.

As graphically captured in Fig. 2.7, there have been both immediate impacts and broader consequences. While the former have manifested themselves in the previously discussed string of launch and spacecraft failures, in the budget reduction, in the delay—if not cancellation—of several programmes as well as in the strains on commercial operations, the latter have been evidenced in important transformations in Russia's institutional setting for space, in its policy orientations, and in its programmes and technological development. More specifically:

- At institutional level, the most important impact has been the acceleration and change of directions in the industrial overhaul initiated in the mid-2000s and the progressive emergence of a new type of relations with the private space industry.
- On budget and policies, the most significant impact has been a fundamental re-orientation of Russia's space priorities towards more programmatic endeavours as a way to cope with the new budgetary realities.
- At programmatic level, the most notable impact has been the delay and cancellation of several programmes (see Annex C for an overview of this rescheduling) as well as an increasing effort to reach independence for critical technologies.

These broader implications will be assessed in further detail in the following sections.

2.2   Evolution and Outlook

This section provides an assessment of the current Russian space programme. In the first part, the institutional framework for space activities in Russia is analysed by reviewing its historical evolution, the current setting and the main actors. The second section provides an overview of the budget and policies, and the third deals with Russia's capabilities and programmes.

2.2.1 The Institutional and Industrial Setting: Historical Evolution and Current Landscape

The Russian space programme has never been separated from the government, and all its activities (including commercial operations and external cooperation) have been executed through Roscosmos and its predecessors. The significant role played by the Russian government is exemplified by the fact that a private space sector as understood in the western sense hardly exists in Russia.[12] In fact, when looking at the industrial entities carrying out space activities in Russia, full state ownership has been put in place as a standard.

2.2.1.1 The Evolution of Russia's Space Orgware: 1992–2017

Changes in the organisation of Russian space activities have been almost constant over the past 25 years. Following the dissolution of the ministerial structures under the USSR regime, the Russian Space Agency (RKA) was established in 1992 with the aim of coping with the new institutional reality and relative needs of post-Soviet Russia. RKA was mainly responsible for the management and implementation of civil space activities, while military space activities were the responsibility of the Military Space Forces.

In 1999, the RKA was restructured to encompass also aeronautical activities, and its official name was changed to *Rosaviacosmos*. However, the reorganization of the Russian space agency into an aeronautical and space administration eventually failed due to an almost innate difference between Russian aviation and space industry (unlike other countries, the country's space industry essentially originated from munitions and artillery industries, rather than from aviation).

Therefore, in 2004, the aeronautical and space activities were split again and the space agency became the Federal Space Agency (FSA) or *Roscosmos*. Anatoly N. Perminov replaced Yuri N. Koptev as the head of the agency. Following the governmental decree "On the systems and structure of federal bodies representing executive power" of 9 March 2004, Roscosmos gained ministerial status as it became one of the 28 Federal Agencies, reporting directly to the Government.

---

[12]Even in the telecommunications sector timid trends towards an increased private sector role, as in Gazprom's satellite business, are overshadowed by government control as the satellites are part of the responsibility of the Ministry of Telecommunication and Mass Communication, thereby imposing the use of Russian launch services.

From 2004 onwards, Roscosmos thus operated as an independent authority of executive power in charge of civil space activities as well as a hub for coordinating military space activities with the Ministry of Defence. Its main responsibilities included:

- Implementing the space policy developed by the government
- Implementing the Federal Space Programme and coordinating the work conducted at the Baikonur spaceport
- Managing the state property in space infrastructure
- Coordinating international cooperation (including for commercial ventures).

The Agency was governed by its head and five deputies plus eight departments and several offices. At that time, the agency directly employed 300 people, though it also exercised administrative control over research and production companies that formed the backbone of Russia's space sector. The FSA was more specifically responsible for 112 organizations, which represents most of the Russian space sector. Throughout the 2000s, these enterprises employed 250,000 people and were divided into:

- Design offices, scientific research and design institutes
- Joint-Stock Industrial companies (JSC)
- Federal State Unitary enterprises (FGUP).[13]

In order to streamline the relations between these constituent organisations and enhance their competitiveness, in January 2006 the Government ratified the "Strategy for Development of the Space Industry up to 2015", through which a restriction of the space industry was initiated. The strategy included the formation of a new organisational structure of the industrial base by creating 10 horizontally and vertically integrated structures by 2010, and setting up 3–4 space corporations that would include most of the main enterprises in the field by 2015. The restructuring mainly took place around:

- the Khrunichev State Research and Production Space Centre (KhSC), which was reorganised around launcher activities,[14]
- the Reshetnev Applied Mechanics Research and Production Association (NPO PM), which was reorganised around communications and navigation satellite activities,[15]

---

[13]The concept of FGUPs emerged with the beginning of the market reforms in the mid-1990s. The driver was to secure and protect sensitive government property, e.g. enterprises of the military industrial complex, in the turmoil of several privatisation campaigns. In 2002, the status of unitary enterprise was regulated by Federal Law. For space FGUPs, the manager is appointed by Roscosmos as authorised designated entity, to which the FGUP also reports. The other major organisational form is the JSCs, typically with 100% of the stock owned by the government.

[14]KhSC was reorganised around the launcher activities according to a Presidential Decree of January 2007. The four subsidiaries of KhSC are: Polyot Production Association; Voronezh Mechanical Plant; the Moscow Company "Dlina"; and the Isaev Design Bureau for Chemical Engineering.

[15]On 9 June 2006, the President of the Russian Federation signed the Decree "Related the Joint-Stock Company Information Satellite Systems" to be established under the responsibility of Reshetnev Applied Mechanics Research and Production Association NPO PM. The subsidiaries of the

- the Russian Research Institute for Space Instrument Engineering (RNII KP), which was restructured together with NII TP, NII FI, NPO IT, NII KP NPO OrionOKB MEI.
- the Glushko Energomash Research and Production Association (Energomash NPO), which was reorganised around the field of propulsion,
- the Reutov Engineering Research and Production Association in the field of electronics.[16]

The first part of the restructuring process was completed by 2010. Following the string of launch failures that began in 2010 and the constant delays in the deployment of several assets despite the increasing funding, it was decided to kick-start an even more comprehensive reform of the Russian space industry (Moskowitz, 2011).

As a first step, the top management of Roscosmos was changed in 2011 in the hope of strengthening financial control measures and improving quality control by creating several new departments including a Quality Management Department. However, the subsequent launch failures between August 2011 and December 2012 underscored the necessity of deeper reforms. In response to the request of the Kremlin to deliver proposals for the reorganisation of the sector, two controversial solutions were advanced:

- the creation of a state corporation comprising both industrial and public functions (following the example of the Rosatom State Corporation, which was established in 2007 to manage Russian nuclear activities)
- the merging of all space industries into one state-owned holding separated from the state agency (with a procurement model similar to that in Europe).

The 2013–2014 Reform

With the successive launch failures in 2012/2013 the lack of a separation line between customer and contractor functions was identified as the major cause of the decreased industrial quality, and preference was thus given, at least initially, to the second option. This resulted in Russian President Vladimir Putin signing a decree in December 2013 that approved the plan of uniting the industrial leaders of space industry while separating agency and contractor functions. Oleg Ostapenko was appointed head of Roscosmos to guide this reform with the overall objective of streamlining production and operation of spacecraft and cutting down on the misuse of funds (Astrowatch, 2014).

---

new enterprise are: Research & Production Center "Polyus" (Tomsk); the Research & Production Enterprise "Kvant" (Moscow); Siberian Devices and Systems (Omsk); Research & Production Enterprise "Geofizika-Cosmos" (Moscow); Research & Production Enterprise of Space Instrument-making (Rostov-on-Don); Sibpromproekt (Zheleznogorsk); NPO PM—Razvitie (Zheleznogorsk); NPO PM—Small Design Bureau and Testing Technical Center (Zheleznogorsk).

[16] A specific case is represented by RSC Energia, which was restructured earlier into a joint-stock company much earlier, in 1995 following the Presidential Decree No. 237 "On the Privatization Procedure for the Scientific-Production Association Energia after Academician S.P. Korolev" of 4 February 1994 and based upon the Government Decision No. 415 "On Creating S.P. Korolev Rocket-Space Corporation Energia" of 29 April 1994 (RSC Energia, 2018). In addition, RSC Energia remained a relatively independent space corporation as compared to the other research and production organisations.

The reform first split the Federal Space Agency Roscosmos in two, a demand and a supply side, where Roscosmos was intended to act as a customer, responsible for strategic planning, space policy, research and ground infrastructure—including the cosmodromes (Reuters, 2013), while the supply side consisted of a holding company consolidating most of the sector's companies developing and manufacturing spacecraft into the United Rocket and Space Corporation (URSC) (TASS News Agency, 2013).

The United Rocket and Space Corporation (URSC) was officially established in April 2014 and Igor Komarov—former manager of AvtoVAZ, Russia's biggest automobile enterprise—became its head. According to the reform plans, URSC was to regroup about 62 companies including all major launch vehicle and satellite developers and manufacturers, except for 11 entities remaining under Roscosmos's direct control (e.g. launch site operator and launch service provider TsENKI, and the Cosmonauts Training Centre). By 2015, the published URSC structure included 10 integrated structures (including Khrunichev, Makeyev, Russian Space Systems, Energia and Progress) regrouping some 40 companies and 14 independent organisations (8 JSCs and 6 FGUPs) 2017 (International Astronautical Federation, 2014).

During the process, the responsibilities of URSC were also expanded, thus complicating relations with the Federal Space Agency. Indeed, although certain tasks were initially excluded from the URSC portfolio, in December 2014, an executive order conferred URSC with important prerogatives. In accordance with this executive order, URSC was to provide advice and be involved in the drafting of defence strategic planning. This was formerly an exclusive prerogative of Roscosmos. The executive order permitted URSC to propose to the Russian Government candidates to be appointed to top management positions for Russian enterprises and corporations—again a former prerogative of Roscosmos. Furthermore, URSC could exercise management and control of the state-owned space enterprises until their conversion to public corporations. It may be speculated that this executive misalignment was the by-product of a power play between Igor Komarov, head of URSC, and Oleg Ostapenko, head of Roscosmos.

The 2015–2016 Reforms

Although the overarching idea of the reform initiated in 2013/2014 was that Roscosmos would stay in charge of strategic planning and administrative work while the United Rocket and Space Corporation (URSC) would subsume the Russian space industry, it soon became clear that this approach was not functional, owing to the progressive duplication of bureaucracy within the two organisations and constant misunderstanding—not to say disputes—over their prerogatives and functions.

In an apparent 360° change of course, it was then decided to switch back to the first proposal and to restructure Roscosmos into a State Corporation to combine all the possible functions under one roof (Zhukov, 2015). On 21 January 2015, Oleg Ostapenko was replaced by Igor Komarov as the Head of Roscosmos. On the same day, President Putin approved the merging of Roscosmos's duties with the URSC, establishing a single entity "*Roscosmos State Corporation*" (RSC) at the helm of Russia's space industry (Bodner, 2015).

Fig. 2.8 Russian space agency timeline

The form of a state corporation was chosen following the example of Rosatom SC (which encompasses over 300 companies operating in the nuclear energy sector) with the aim of avoiding doubling functionalities as well as legal problems regarding responsibilities for state owned property.[17] In addition, by consolidating the industry, the goal was:

- to decrease budget expenditure by eliminating excess manufacturing capacity in both infrastructure and workforce;
- to better address quality control issues;
- to streamline the procurement of foreign components with increased purchasing power to negotiate volume-based discounts (Space News, 2013).

Probably the main outcome of these changes is the gradual liquidation of FGUPs, turning them into public joint-stock companies that are more transparent and independent. They are obliged to publish annual reports and have a right to sell their territories (thus the Khrunichev State Research and Production Space Centre can lose 80% of its initial territory due to the optimisation measures).

The reform was eventually approved by the State Duma on 1 July 2015, and on 13 July President Putin ratified the law governing the new Roscosmos structure, which combined the functions of policy making, strategic planning and managing the entire rocket and space industry in the country.

Thus, on 1 January 2016 the Federal Space Agency was dissolved and its responsibilities were transferred to Roscosmos State Corporation. As for the URSC, it continued to exist as a sub-entity of Roscosmos SC, with the deputy director of this SC being at the same time the Director General of URSC.

The reform process of the Russian space sector was completed by December 2017. Its 25 year-long evolution is captured in Fig. 2.8.

2.2.1.2 The Current Institutional and Industrial Setting: Key Organisations and Human Capacity

Roscosmos Space Corporation

As a result of this consolidation, current Russian space activities are centred on Roscosmos Space Corporation (RSC), which now comprises and oversees all Russian space-related entities, namely:

---

[17] A State Corporation (SC) is a non-profit organisation subject to reduced control and supervision from government bodies, weaker requirements on information disclosure and immunity against bankruptcy.

(a) The former Roscosmos federal space agency;
(b) The United Rocket and Space Corporation, which itself incorporates all the
     major industrial organisations in the country such as:

- *S.P. Korolev RKK Energia*—the manufacturer of upper stages, Progress and
  Soyuz spacecraft and satellite platform Yamal; it also operates the Russian
  ISS segment and the Sea Launch programme;
- *RSC Progress*, which is responsible for Soyuz launch vehicles;
- The Khrunichev State Research and Production Space Centre, that back in
  1993 integrated all of its primary subcontractors and developed and produced
  Proton, Rockot and Angara launch vehicles;
- *NPO Energomash*, that is in charge of the development of engines for Cosmos,
  Proton, Angara and Zenit launchers, and RD-180 engines in cooperation with
  Pratt & Whitney;
- The *Moscow Institute of Thermal Technologies*;
- *NPO Lavochkin* that works on satellites manufacturing and produces the Fre-
  gat upper stage;
- *Fakel Design Bureau* (electric propulsion);
- *Arsenal Machine Building Plant*;
- *Information Satellite Systems Reshetnev* (communications satellite manufac-
  turer);

(c) Other institutions under control of the former Federal Space Agency such as:

- the Cosmonauts Training Centre (CTC),
- the TsENKI Centre for Operation of Space Ground-Based Infrastructure and
  all the design facilities, operators, vendors and suppliers that have been centralised under it,
- Glavkosmos Launch Services, which provides support to Roscosmos international contracts and activities.

The new Roscosmos State Corporation is organised around eight holdings spe-
cialized in various fields of rocketry and space flight, including: (a) *human space
flights*, (b) *launch systems*, (c) *unmanned spacecraft*, (d) *rocket propulsion*, (e) *military missiles*, (f) *space avionics*, (g) *special military space systems*, and, (h) *flight
control systems*. Such a division of responsibilities follows the lines of traditional
specialization among key companies in the—Russian space industry.

Industrial Infrastructure Optimisation and Human Capacity

One of the main objectives behind the 2015 institutional overhaul was to respond
to the need for an improved economic performance of the Russian space activities
under tightening financial possibilities. In an interview during the 2016 Economic
Forum, Igor Komarov confirmed that the streamlining and optimisation of both the
infrastructure and workforce were the major instruments that the newly created RSC
intended to use to tackle this issue (IZ Russia, 2016).

As already highlighted in Sect. 2.1.1, Russia's space industry's capacity has been
under-utilized since the 1990s, owing to the mismatch between the level of ambition

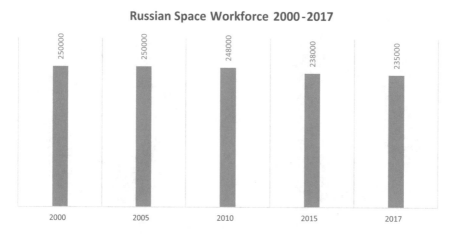

Fig. 2.9 The evolution of Russia's space workforce

of the post-Soviet space programme and the large infrastructure and workforce put in place under the USSR legacy. Capacity utilisation was estimated to be 52.4% in 2013, 49.9% in 2014 and 50.1% in 2015.

While the streamlining of the operational facilities has been facilitated by the recent overhaul, workforce optimisation has proved somehow more cumbersome. Over the past 15 years, the overall workforce in the Russian space sector has decreased by approximately 15,000 units. While this reduction appears significant in absolute terms, it still appears modest when compared to the optimised number of 150,000 employees estimated by Roscosmos (Fig 2.9).

According to a special report by RBC issued in April 2017 (Homchenko, 2017), 235,000 people are currently employed in the Russian space sector. More than 80% (196,000 people) of the employees work for the entities within the URSC, while the remaining 20% are employed in other institutes such as the CTC) and the TsENKI Centre. While no updated information has been provided concerning the typology of the workforce, it is known that the 2014 allocation of occupations was the following: 22%—scientists and high-level specialists, 15%—managers, 63%—workmen (see Fig. 2.10).

As for the age cohorts, in 2015 almost half of the personnel were over 50 years old, while the share of young specialists was less than 23%. Considering that Russia's specialist population is approaching retirement, the space sector will need to find a skilled and young workforce, the lack of which is today one of the major causes behind the crisis of the Russian space sector (see Sect. 2.1.1). As recently confirmed by RKK Energia's secretary Denis Kravchenko, over the next 10 years the space industry will need at least 100,000 highly qualified specialists. These figures correspond to approximately the 42% of the current workforce.

Although recent data published by Russia's Ministry of Education show that the demand for admission in engineering faculties in the country has been steadily

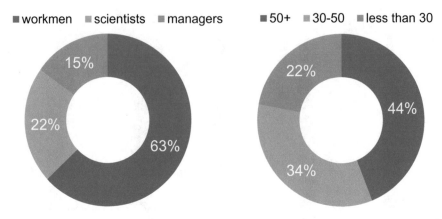

Fig. 2.10 Human resource breakdown (by typology and age Cohort)

growing over the last few years, there are still several factors that negatively impact the popularity of space-related curricula and careers. The most important one is, of course, the low remuneration level in the sector. According to interviews, in 2016 the average salary was 46,600 RUB, corresponding to roughly €650 per month. Another major disincentive for young Russians to join the sector are the restrictions imposed on the sector's workforce for travelling abroad. It has been reported that even students enrolled in aerospace university courses need to undergo cumbersome procedures to obtain passports and visas (Cherebko, 2016).

Against this background, the government has recently started to support space industry through a series of outreach measures. Among them, particularly noteworthy is that in 2017, for the second time in Soviet/Russian history, the recruitment to the CTC Cosmonaut Corps was open to every citizen meeting basic requirements so as to make space more accessible—or at least more popular—among Russian citizens. In addition, Roscosmos has recently created a website (http://keystart.roscosmos.ru) which aims to promote Science, Technology, Engineering and Mathematics (STEM) disciplines and encourage students to choose their career in the space industry. Finally, two movies have been filmed with the support of the government to draw citizens' pride from their country's glorious space legacy. They were released in 2017 in conjunction with the Sputnik launch celebrations.

The Russian Academy of Sciences

In addition to Roscosmos SC, another key actor in the current organisational setting of Russian space activities is the Russian Academy of Sciences (RAS), which plays an important role particularly as concerns space sciences and exploration. The RAS was first established in 1724 and remained a key institution during the Soviet Union. Its importance was further reaffirmed when it was reinstated as the supreme scientific institution of Russia by a Decree of the President in 1991. However, during the 1990s, the RAS witness a relative decline because of the radical reduction of the funding made available by post-Soviet leaders. As pointed out by

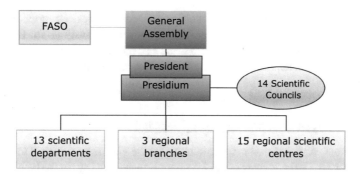

Fig. 2.11 Structure of the Russian academy of sciences

Tatiana Stanovaya, "this trend was reversed over the past decade, when RAS's budget was significantly increased, eventually reaching 64 billion roubles in 2012" (€ 1.68 billion) (Stanovaya, 2013). Several problems, such as the "lack of dynamic development, the advanced age of academics (65 years), corruption, poor financial management and a significant decline in scientific publications" remained, however, unaddressed (Ibid).

In light of this backdrop, in 2013, the RAS underwent a major reform process primarily aiming to increase the participation of universities in research and to better oversee research priorities and spending. While this reform was long-awaited, it was hastily received by the scientific community, *inter alia* because it placed all Russian academic and scientific institutions under the jurisdiction of a special federal executive body, the newly created Federal Agency for Scientific Organisations (FASO), thereby sensibly reducing the Academy's independence (Clark, 2013). In addition, the reforms defined a new procedure for the appointment of the various institutions' directors, with the candidates proposed by RAS requiring the approval of FASO. Despite the lukewarm reception and even public protests, in September 2013 President Putin approved the federal law "On the Russian Academy of Sciences and the Decree "On the Federal Agency for Scientific Organisations".[18]

As shown in Fig. 2.11, the RAS currently reports to the FASO. It includes 13 scientific departments (structured by scientific field), 3 regional branches (the Siberian Branch, the Ural Branch and the Far East Branch), and 15 regional scientific centres. In 2017, the RAS employed more than 90,000 people, including more than 48,000 scientific staff. As with Roscosmos, over the past 15 years the tendency has been to gradually reduce this large number of staff.

Space-related activities are conducted mainly in the Department of Physical Sciences within its General Physics and Astronomy Section, in particular in the Space Research Institute (SRI or IKI in Russian), as well as in the Institute of Astronomy, the Pulkovo Observatory, the Institute of Applied Astronomy, the Institute of

---

[18]Three years later, the intermediate results of the reform have been estimated negatively by the scientific community (McGilvray, 2016).

Radio Engineering and Electronics, the Special Astrophysical Observatory, and in the Department of Physical Chemistry and Biology, in the Institute for Biomedical Problems (IBMP) of the State Research Centre of the Russian Federation.

Among the 14 Scientific Councils of the RAS Presidium, there is a Space Council. In addition, an "Interdepartmental Council on Space Power Problems" was established between the RAS and Roscosmos.

The Ministry of Defence and the Space Command

A traditional feature of Russian space activities is the high involvement of the military. The Ministry of Defence shares with Roscosmos responsibility for the Federal Target Programme on cosmodromes and has a large responsibility in the modernisation and management of the GLONASS dual-use system. It is responsible for two of the five sub-programmes of the GLONASS programme (see Sect. 2.2.2). The Ministry is responsible for important military space programmes, covering a wide spectrum and including signals and electronic intelligence, and reconnaissance missions and military launches.

Another key player on the military side is the Military Space Command (formerly called Space Command, Space Forces or VKS), which is part of the Aerospace Defence Forces. The Space Command is in charge of military space operations as well as of a large part of launch site operations throughout Russian territory, a responsibility shared with Roscosmos. The Aerospace Defence Forces comprise four major elements:

- the Space Command,
- the Air Defence and Anti-Ballistic Missile Command,
- Plesetsk Cosmodrome,
- the Arsenal—Machine Building Plant.

Skolkovo Space Cluster and Russian Private Industry

Notwithstanding the recent industrial consolidation within the RSC framework, over the past few years the Russian government has undertaken some efforts to make its space sector open to involvement by private industry and hence increase its overall competitiveness, both domestically and internationally. Acknowledging that in Russia's context the possibility of private enterprises and start-ups competing for space procurement or entering the value chains of the bigger companies was only theoretical without the creation of an effective private ecosystem, in 2010, Russian President Medvedev launched the Skolkovo Foundation, which included a Space and Telecommunications "cluster" among its core activities.

Partially inspired by the Silicon Valley's innovation models and entrepreneurship, the overarching goal of the Skolkovo Foundation is to provide catalysts for the diversification of the Russian economy, by creating a sustainable ecosystem of entrepreneurship and innovation, stimulating a start-up culture, and encouraging venture capitalism. In addition to space technologies, Skolkovo Foundation focuses on other technology domains, identified as "growth points" by the Russian expert community: energy efficiency, strategic computer technologies, biomedicine, and nuclear technologies.

The most striking feature of Skolkovo is that it is the only development institution supporting small private space businesses as a primary focus. Indeed, current clusters in Russia focus on other technological domains or support fundamental scientific research rather than high-tech business (Kosenkov 2016). To support the growth of a vibrant ecosystem for private space companies, over the past seven years Skolkovo Space cluster has undertaken multiple actions in various directions. Three are particularly noteworthy.

- First, Skolkovo has been providing small companies with favourable conditions for pursuing their business, including tax benefits, access to venture investment, technologies from industrial partners, small grant funding, and expertise with respect to both space technologies and business development.[19] The space cluster has intentionally positioned itself as a point of attraction for building up a strong community of entrepreneurs, experts, scientists, investors, and a platform for dialogue between public and private players with the aim of facilitating the identification of mutually beneficial interests (Ibid).
- Skolkovo has also fostered an efficient collaboration framework with Roscosmos, by identifying favourable conditions for small and medium enterprises' activities in the space domain, particularly with respect to the licenses issued by Roscosmos. The space cluster has also obtained access to industry expertise in project evaluation through cooperation with Roscosmos. These activities have in turn strengthened the space cluster's position as an independent centre of expertise for the elaboration of Russian space policy, particularly of public-private partnerships in space activities.[20] In addition, some of the space cluster's resident companies have gained the possibility to participate Roscosmos programmes and tenders. Vice versa, some of the major Russian public enterprises (e.g. NPO "Energomash" and ISS Reshetnev) have founded spin-off companies in the space cluster focusing on R&D activities.
- Third, taking into account that space start-ups can thrive only if their solutions are offered on the international market, Skolkovo has strengthened cooperation with research institutions and space start-ups worldwide, by establishing close connections with science and technology parks, hosting space start-ups, proposing them a soft-landing in Skolkovo and providing them with grant funding and acceleration for Russian market access. The collaboration between Skolkovo resident companies and foreign companies has been intended as a tool to support the possible establishment of globally competitive joint ventures. Notably, Skolkovo

---

[19]The domains supported by the cluster are: Satellite applications, including satellite navigation and Earth observation; Small spacecraft technologies; Spacecraft systems, subsystems and components; Ground infrastructure for space activities; Industrial and manufacturing technologies for aerospace industry; New materials for aerospace applications; Orbital and suborbital launch systems; Unmanned aerial vehicles and its applications; New telecom technologies, including those for wired and wireless telecommunications; Software and hardware for telecommunications.

[20]Skolkovo experts have also made significant contributions to the reform of space sector by bringing new approaches to the general vision elaborated by experts.

has also encouraged partnerships with international industry giants such as Airbus and Boeing, offering the possibility of establishing some of their research centres in Skolkovo. The overarching idea is that leading aerospace manufacturers could in the future become strategic investors for some of the spin-off companies (Ibid).

The result of Skolkovo cluster activities is the emergence of an ecosystem of more than 150 small and medium companies in various technological domains related to space: from development of launch vehicles and satellites to new applications and products using space data. Skolkovo's support has enabled them to find investments, partners, and clients on the domestic and international markets. A list of Russia's most active private players is provided in Table 2.4.

After seven years of activity, Skolkovo has certainly many laudable achievements to look back upon, but it is also evident that its resident companies continue to face numerous challenges in moving towards sustainable growth.

For one thing, the entry barrier for space businesses in Russia remain very high and it is difficult for start-ups to integrate with the space industry value chains, especially in light of the recent institutional reform of the sector. "The customers, contractors, and even the regulator are still experiencing the turbulence of the recent mergers and the sector governance is still not properly established. Only in 2016 Roscosmos said it would allow private companies access to the space services market, and not before 2020. In addition, many report "passive resistance from Roscosmos against private companies, for example, demanding detailed designs and models of proposed systems before discussing funding. This is not surprising since as a state corporation, Roscosmos does not have much reason to support private start-ups that become competitors" (McClintock, 2017).

At the same time, with the exception of large companies like Gazprom and S7 (see further), most of Russian venture investors remains reluctant to participate in risky space projects, which demand massive cash infusion during long periods. Also, "the number of investors having the sufficient expertise in space business remains relatively low" (Kosenkov, 2016), and some point out that "Russians, often capable of great technological innovation, are not as steeped in the capitalist ethos of recognizing and addressing needs of the market" (McClintock, 2017).

S7 Space Transportation Systems

The Novosibirsk-based S7 Group, which primarily operates Siberian Airlines, represents a particularly special case in the current landscape of Russian space activities. The company joined the limited number of private Russian companies investing in commercial space in September 2016 after signing an agreement with the Sea Launch Group for the acquisition of all its assets (including two vessels and the infrastructure located in California) for $160 million. The S7 Space Transportation Systems, as part of the S7 Group, received the governmental license to engage in space activities and in international operations in February 2017 and is currently preparing for commercialising both the Zenit-3SL and Soyuz-5 rocket, which is under development by Roscosmos SC, on the international market. The company's CEO, Sergey Sopov, has announced that the deal can be finalised as early as April 2018. A more detailed overview of the evolution of Sea Launch is provided in Annex D.

Table 2.4 Russia's new space ecosystem

Company	Segment	Remarks
Dauria Aerospace	EO small satellites	The company built and launched two Perseus-M micro-satellites in the U.S. in 2014. These maritime surveillance satellites provide automatic ship identification and serve as technological test beds for future platforms and equipment. The company is currently developing and manufacturing small and middle-sized spacecraft for remote sensing and telecommunications
Sputnix	Micro-satellites	The company developed a micro satellite, TabletSat-Aurora, which was the first private satellite built and launched in Russia in 2014. The novelty of the project is that all the platform's service systems include a common service interface, which permits building the satellites from standardized blocks. Sputnix aims to make TabletSat-Aurora a universal platform for different payloads. The company also plans to build an EO constellation of small satellites providing high-resolution imagery
Spectralaser	Engines components	The company offers laser ignition modules for rocket engines. It has successfully tested the product on liquid-fuel rocket engines of the Soyuz launch vehicle. The research has been conducted in close cooperation with several industrial partners, including NPO Energomash, JSC RSC Progress and Kuznetsov JSC
KosmoKurs	Suborbital launch vehicle	The company is developing a reusable suborbital launch vehicle for tourism and scientific experiments. The system includes a multi-level emergency protection system ensuring safe and low-cost suborbital flight for passengers. After initial operations, it plans to develop launch vehicles for small satellites based on the system. The stage of flight proven prototype is supposed to be achieved by 2020
NSTR Rocket Technologies	Micro-launcher	The company is developing ERRAI—the ultra-light rocket for launching nano-satellites with a weight of 1–10 kg. It also works on AstroNYX—a network of automated telescopes that will be operated by the mobile app or from the browser. The company operates outside of the Skolkovo space cluster
Lin Industrial	Micro-launcher	Lin Industrial is developing a family of light launch vehicles for small satellite launches with the goal of drastically reducing the costs of accessing space. As of 2017, they were still working on the rocket engines
WayRay	Applications	Selected by Forbes as among "The 6 Tech Companies Disrupting the Daily Drive", WayRay offers augmented reality technology for vehicles using universal windshield coating combining the features of smartphone, navigation device and head-up display
SPIRIT Navigation	Navigation technology	The company develops hybrid navigation technology for seamless positioning indoors and outdoors. The core product is a hypersensitive software for GLONASS-GPS signal reception, ensuring sensitivity 5 times higher than world market leaders

(continued)

Table 2.4 (continued)

Company	Segment	Remarks
RoboCV	Robotic equipment	RoboCV has developed an autopilot system for warehouse equipment, X-MOTION, using its research on the Moon rover for the Google X-Prize competition. The product has been tested on Samsung, Volkswagen, and Magnet warehouses. The company has raised more than 100 million RUB of venture financing
SUKhE	Energy storage	The company has developed automated optimization systems for energy-efficient use of autonomous power sources for transport and energy applications. Initially designed for the ground segment of space systems, the system is now being successfully tested to be applied on new electronic buses developed in Saratov

2.2.2   Space Policies and Funding

As regulated by the law "On Space Activities", adopted in August 1993 and amended several times over the past 20 years, the main stakeholders in the Russian space policy-making process are the President, the Government (in particular the Ministry of Defence and the Ministry of Finance), and Roscosmos State Corporation.

The President of the Russian Federation exercises overall leadership of space activities. The president asserts the main principles of the Russian space policy through presidential decrees. The government examines and approves Russian space-related programmes. The Security Council, as a department of the Presidential Executive Office, advises the president on space issues related to national security (see Fig. 2.12).

Space activities in the Russian Federation are conducted on the basis of various long-term Federal Target Programmes (FTP) and Sub-Programmes (SB). These programmes are funded to a large extent by the federal budget but also by non-budgetary governmental funds (which includes funding from commercial launches of foreign cosmonauts, from joint ventures with foreign partners, and Russian participation in international projects). The federal budget is adopted by the State Duma and approved by the Federation Council.

Given the inherently stove-piped nature of the FTPs and SBs, reform of the policy on definition of space programmes was implemented in 2012, through which the existing space programmes and sub-programmes were consolidated under a co-management scheme between Roscosmos, acting as executive managing body, and the Ministry of Defence.[21]

---

[21] The reform was implemented with the "Space Activities of Russia in 2013–2020" programme, which was signed by Presidential Decree No. 2594 on 28 December 2012. It foresees two implementation phases: 2013–2015 (implementation of the current FSP 2015) and 2016-2020 (in line with the "Development Strategy of the Rocket and Space Industry through the year 2013 and beyond"), and financing of approximately RUB2.1 trillion over the period.

Fig. 2.12 Space
policy-making in Russia

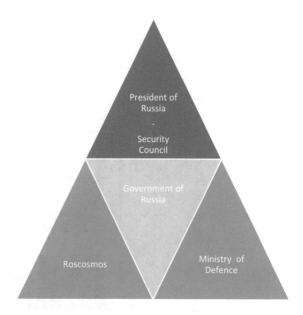

This co-management scheme—as implemented through the Development Strategy of the Rocket and Space Industry through the year 2013 and beyond—currently comprises five major programmes, namely three FTPs and two SBs:

- The FTP on Development, Utilisation and Maintenance of the GLONASS System for 2012–2020 (GLONASS-2020)
- The FTP on Development of Russia's Cosmodromes for 2017–2025
- The FTP on the Federal Space Programme of Russia for 2016–2025 (FSP-2025)
- The SB on Support of the State Programme Implementation
- The SB on Priority Innovation Projects of Rocket and Space Industry (Fig. 2.13).

2.2.2.1 The Federal Space Programme 2016–2025

The Federal Space Programme 2016–2025 (FSP-2025) is by far the most important of these five elements as it comprises more substantial programmes and a larger part of budget as compared to the two other FTPs and SBs.

Proposing the Federal Space Programme is mainly entrusted to Roscosmos. The programmatic part is typically drafted by Roscosmos SC in concordance with other stakeholders (e.g. the Space Council of the Russian Academy of Sciences, the Ministry of Construction, the Ministry of Defence, etc.), before submission to the Government.

Concerning the budgetary part of the FSP, it is mainly defined by Roscosmos and the Ministry of Finance covering the entire period before the formal approval of the programme. As implementing a long-term programme clearly imposes some

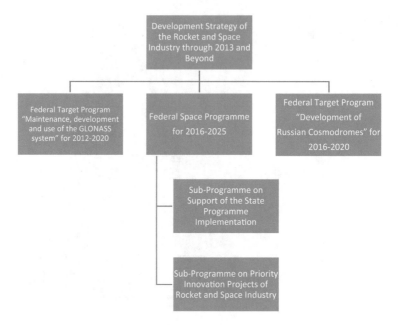

Fig. 2.13 Russia's federal space programmes

budgetary flexibility, since 2008 a new budgetary principle has been implemented that provides for a sliding three-year plan with committed budgetary figures for the first year and re-adjustable benchmark figures for the two following years, allowing for yearly reallocations and readjustments for inflation.

The FSP-2025 is the fourth federal space programme approved by the Russian Federation and covers the period 2016–2025. The three previous space programmes took place over the 1992–2000, the 2001–2005 and the 2006–2015 periods.

The programme was adopted by Decree of the Government of the Russian Federation on 23 March 2016. The document, however, had been in the making for over two years, with seemingly endless negotiations among various federal entities concerning both the financial and programmatic parts.

As mentioned in Sect. 2.1.2, the total budget of the programme over the 2016–2025 period was initially estimated at about 2849 billion roubles (€56.9 billion).[22] The programme included a number of ambitious projects including, for instance, a super heavy-lift launch vehicle (80 tonnes LEO, upgradeable to 160 tonnes) and the establishment of a manned Moon base by 2020. As a consequence of budget cuts requested by the Russian Finance Ministry in early 2015 to face the challenging economic situation, the initially planned budget was reduced by half in the final version approved in March 2016.

The budget reduction has obviously affected many programmes, causing in particular the abandonment of reusable booster development, the resizing of the super

---

[22] At March 2014 exchange rate 1€ = 50 RUB.

Table 2.5 FSP-2025 budget as approved in March 2016

	Billion Rub	Billion € (March 2016) exchange 1€ = 76 RUB	Billion € (Dec. 2017) exchange 1€ = 69 RUB
2016	104.5	1.37	1.50
2017	104.5	1.37	1.50
2018	104.5	1.37	1.50
2019	117.6	1.54	1.70
2020	136.4	1.79	1.97
2021	155	2.03	2.24
2022	161.9	2.13	2.34
2023	167.9	2.20	2.43
2024	173.9	2.28	2.52
2025	179.7	2.36	2.60
Total 2016–2025	1406	18.50	20.18

heavy-lift launch vehicle development, the reduction of the number of foreseen Angara launch pads in Vostochny from two to one, as well as the cancellation of roughly 30 missions to be performed in the period.[23]

Federal Space Programme Budget

The total budget of the federal space programme over the period 2016–2025 has been set at 1406 billion RUB, corresponding to more than €18 billion (as of December 2017). Table 2.5 provides an overview of the total and annual projected budgets for the FSP-2025. In order to appreciate the fluctuations of the rouble, both the value in Euro at the time of the approval and the value in Euro as of 1 December 2017 are provided.

A graphical illustration of Table 2.5 is provided in Fig. 2.14, which better shows the forecasted trend-line of the budget in the period 2016–2025, and also provides information on the composition of the budget.

When looking at the budget allocations for the FSP-2025, three points must be highlighted:

- The total budget of 1406 billion RUB foreseen for the FSP-2025 has doubled in absolute terms as compared to the previous FSP (2006–2015), which had an adopted budget of 746 billion RUB. However, when considering the exchange rate with Western currencies (US$ or €), the current budget of €1.5 billion per year will be substantially lower than the 2011–2014 period, (when it peaked at the level of roughly €5 billion).
- The annual FSP budget might be different from that contemplated in the original FSP-2025 document due to the above-mentioned revisions and re-allocations that

---

[23]The FSP-2025 initially foresaw a total of 185 launches over the period. After its approval, the FSP currently foresees 155 launches, including manned and unmanned missions.

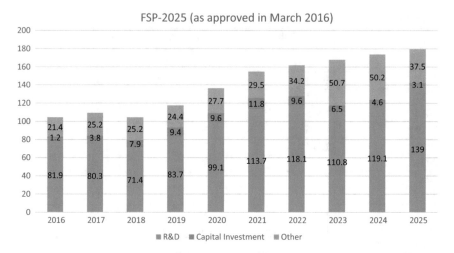

Fig. 2.14 FSP-2025 annual budget composition (in Billion RUB)

take into account the actual funding conditions and status of the projects. In some circumstances, the budget can be increased, while in others reduced. For the period 2017–2019, the government has in fact reduced the annual budget by 14,6% (92,5 billion RUB in 2017, 89,2 billion RUB in 2018, and 86,3 billion RUB in 2019).

• Finally, it should not be overlooked that the FSP represents only one part (albeit the most important one) of the Russian space budget, which is augmented by separate budgets (namely, the FTP on Cosmodromes, the FTP on GLONASS, and the two Sub-programmes). In addition, some 300 billion RUB are expected to be allocated to the FSP-2025 from non-budgetary governmental funds.

Budget Breakdown and Programmatic Priorities

An interesting feature of the FSP 2025 is that it includes the projected budgets for both many individual projects and the different domains of activities. Figure 2.15 provides an overview of the FSP-2025 budget breakdown.

   When looking at the allocation of the budget per activity domain, several programmatic priorities eventually come to the fore. Figure 2.11 more specifically shows that while human spaceflight activities continue to receive the highest budget share (roughly 28% of the total budget), the importance attributed to application missions has significantly grown (EO and telecommunications missions account for more funds than human spaceflight—28% of the FSP budget). This is confirmed by the range of priorities identified by the FSP-2025 document itself, which clearly highlights the effort by the current leadership of Roscosmos to steer the industry toward more pragmatic goals rather than the prestige-oriented projects inherited from the Soviet period. The third place is occupied by launcher programmes (14% of the total budget) while space sciences and exploration is relegated to fifth place (10%), after basic products (11%).

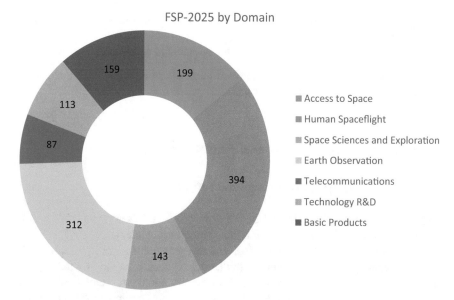

Fig. 2.15 Budget breakdown of the FSP 2025

The programmatic priorities contained in the FSP-2025 and other FTPs are the following

- Launchers:

  - Reduce operational launcher families from 8 to 2 (Angara and Soyuz)
  - Reduce launcher versions from 12 to 6
  - Develop a new medium-lift launcher (Phoenix)
  - Expand Vostochny launch infrastructure

- Human Spaceflight:

  - Support operation of the ISS until at least 2024
  - Launch three Russian modules to the ISS
  - Introduce a new manned launch vehicle "Federation"

- Space Science and Exploration

  - Implement 3 astronomy missions
  - Implement 2 Mars missions
  - Implement 5 lunar missions
  - Develop critical technologies for a manned lunar landing in 2030

- Satellite Systems:

  - Expand the EO constellation from 8 to 23
  - Expand communication satellites from 32 to 41

– Maintain and Expand the GLONASS system from 24 to 30 satellites and improve performance
– Improve the exploitation of data and development of (commercial) applications.

A more detailed description of the main missions and programmes included in the FSP-2025 is provided in the next section.

2.2.3 Current Capabilities and Future Programmes

2.2.3.1 Access to Space: Launchers and Infrastructure

A quarter of century after the fall of the Soviet Union, the position of Russia in the launchers sector is still robust, with the—somewhat challenged—ability to offer commercial launch services at low prices and full state control over the exploitation of its launcher systems. Furthermore, the launchers sector is enjoying the support of a constant and large domestic institutional market, and the retirement of the U.S. Shuttle programme in 2011 gave Russia a temporary monopoly in human access to the ISS. Table 2.6 provides an overview of Russia's current space launch vehicles. A more detailed assessment of the launch vehicles exploited multinationally is in Annex D.

In recent years, Russia has been facing a number of critical issues that threaten its, until-recently, strong presence on the worldwide launchers market. Chief amongst these factors are quality assurance and reliability issues, an increase in prices affected by domestic and non-domestic influences, as well as a perceived reduced autonomy in access to space (see Sect. 2.1).

To overcome these critical issues and cope with evolving demand and increased reliability needs, Russia has concentrated its efforts on the modernization and consolidation of its launcher vehicles, such as the development of the new modular Angara family, formally started already in 1992, while phasing out the Rockot and Dnepr vehicles as the refurbished ICBM stockpile diminishes (Aliberti & Tugnoli, 2016).

The new Angara family has been intended to streamline the manufacturing process, exploit flexibility and modularity, and ultimately provide the capability to launch a complete range of small-to-large payloads. The plans for the Angara family spanned from the small-launcher Angara 1.2 to the heavy lift Angara 7, with Angara 3 (still only at a concept stage) intended to gradually replace Soyuz, and Angara 5 (in an advanced stage of development,) as a future replacement for Proton, post-2020 (Zak, 2015). Both A5 and A7 also foresaw the possibility to be human-rated. In preparation for FSP 2025, a number of ambitious projects were proposed together with Angara, amongst which are the development of a reusable first stage and a new mid-class family of launchers (Phoenix or Soyuz-5), which eventually should replace Soyuz and Zenit.

However, Russia's economic recession, sparked by falling oil prices and the economic sanctions imposed by Western countries, has forced substantial budget cuts to its FSP-2025 launcher programme, which ultimately have led to the scaling down and postponing of the most far-reaching plans. The 2015 cuts more specifically forced the abandonment of reusable booster development, the resizing of heavy lift launch vehicle development (with Angara 7 turning into Angara-5V with the capacity of 35–37 tons to LEO) (Zak, 2017a), as well as a reduction of the number of foreseen Angara launch pads in Vostochny, from two to one (Zyatkov, 2014).

Despite important cuts affecting the number of annual launches and postponing the availability of a heavy lift launcher, the FSP-2025 still envisages funding of 199 billion RUB (roughly €2.6 billion) for launcher activities. These should include both streamlining (by abandoning definitively the Rockot, Dnepr and Zenit programmes) and development activities. In this area, emphasis has been put on further development of Angara 5 (LOX/LH2 stage) as well as on the development of the Phoenix medium launch vehicle up to ground test completion, which should even-

Table 2.6 Russia's current launch vehicles

Launch vehicle	Launch service provider	Performance range (kg)				Launch site
		LEO	SSO	GTO	GEO	
National						
Proton M/Block DM	Khrunichev	19,800		4400	1900	Baikonur
Angara 1.2	Khrunichev	3800				Plesetsk, Vostochny
Angara 5	Khrunichev	24,500		7500	4600	Plesetsk, Vostochny
Soyuz 2-1.a/b	GV Launch Services	7000–8300				Baikonur, Plesetsk Vostochny
Soyuz 2-1.v	TsSKB Progress	2800	2600			Plesetsk, Vostochny
Soyuz FG	TsSKB Progress	7130				Baikonur
(Ex-) Multinational						
Zenit-3SL	Sea Launch/Energia	7300		6200		Odyssey Platform
Zenit-3SLBU	SIS/Yuzhhnoye			4900	1900	Baikonur
Rockot	Eurockot	2100	1600			Plesetsk
Dnepr-1	ISC Kosmotras	3700	2300			Yasny, Baikonur
Proton M	ILS	23,000				Baikonur
Proton M/Breeze M	ILS			6900	3300	Baikonur

Table 2.7 Launch vehicles under development

Launch vehicle	Performance range (Kg)				Launch site	First flight
	LEO	GTO	GEO	Lunar Orbit		
Angara 1.2 (updated)	3000				Plesetsk, Vostochny	>2019
Angara A5P	18,000	6600	4000		Plesetsk, Vostochny	>2019
Angara A5V (proposed)	40,000	13,300	7600		Plesetsk, Vostochny	>n.a
Soyuz-5	17,000	4500	2300		Baikonur, Sea Launch	>2022
Energia-5VK (proposed)	105,000	43,300		20,500	Vostochny	>2028

tually replace the Soyuz family and may incidentally serve as a module of the future heavy launcher.[24] The programme further envisages the development of propulsion technologies (including a methane engine, new liquid propellant engines and basic blocks for engines based on composite materials). Table 2.7 provides an overview of the launchers activities within FSP-2025.

In 2016 the Russian authorities finally gave up the plan for Angara-A5V and started the new booster project called Soyuz-5 (also known as Phoenix in Russian or Sunkar in Kazakh) (Interfax, 2017). In April 2017, the head of RKK Energia was reported to affirm that this rocket will be used for the new-generation piloted spacecraft PTK Federation instead of Angara-5 due to the lower costs (TASS News Agency, 2017). According to Solntsev, the new rocket will be cheaper than the Ukrainian Zenit and American Falcon-9. As announced by the head of Roscosmos Igor Komarov at the 2017 Economic Forum in St-Petersburg, the estimated launch price will be $55 million instead of $70 million for Zenit (Brjanski, 2017).

Originally intended as a Russian-Kazakh vehicle for the commercial satellite market, Phoenix turned into a new RKK Energia project for manned space flight and a new potential client for the Baikonur cosmodrome, where the Russia–Kazakhstan Baiterek JV is currently under development. On May 2017, President Putin suggested constructing the base for the PTK Federation in Baikonur instead of Vostochny, despite all the previous plans, due to the switch from Angara-5 to Soyuz-5. According to Dmitry Rogozin, the rocket shall be ready by 2021. Soyuz-5, which has a payload capacity of 17 tons to LEO, is now meant to launch the Russian segments of the ISS and the first crew in 2024.

In June 2017 Russia returned to the idea of building a super heavy-lift rocket based on joining together several Soyuz-5 stages. Considering the number of plans in which Soyuz-5 is now appearing it can be called the most prospective Russian vehicle for the next 2 decades, along with Proton-M that was returned to the operating state in 2017.

---

[24]The Phoenix project is linked to the joint Baiterek project with Kazakhstan (see Sect. 3.2).

Table 2.8 Russia's launch infrastructure

Name	Location	Operator	Main launch vehicle
Plesetsk Cosmodrome	Mirnyi, Arkhangelsk, Russia	RSC	Angara, Soyuz, Cosmos, Cyclone 3, Rockot, Soyuz 2-1.v, Start-1
Baikonur Cosmodrome	Kzyl-Ordinsk, Kazakhstan	RSC	Proton, Soyuz, Cosmos, Cyclone 2, Dnepr, Strela
Vostochny Cosmodrome	Amurskaya, Russia	RSC	All Russian upcoming manned and unmanned LV
Sea Launch Platform	Long Beach, CA, United States	S7 Group	Zenit (assembled in the U.S.) from 2019, Soyuz-5 from 2022
Yasny	Dombarovsky, Orenburg, Russia	RSC	Dnepr
Barents Launch Area	Barents Sea, Russia	VMF	Shtil, Volna

The new super-heavy lift launcher, which has received the designation of Energia-5 kV, is primarily intended for future human missions to the Moon. The payload could include the piloted spacecraft "Federation", a lunar lander, or a lunar cargo vehicle. The design proposed by RKK Energia currently envisages the development of a five-stage launcher, with a lift-off mass of 2300 tons and a payload capacity of 105 tons to LEO and 20.5 tons to lunar orbit (Zak, 2017b).

Launch Infrastructure

Russian launch infrastructures, including test facilities and launch complexes, are state-owned and have traditionally been under the control of the Aerospace Defence Forces. However, following a domestic space transport restructuring process, the responsibility for the Baikonur infrastructure was transferred to Roscosmos in 2007, with Baikonur thereby becoming a fully civilian spaceport. On the other hand, Plesetsk facilities remained under the control and supervision of the Aerospace Defence Forces. Russia's current launch infrastructure is comprised of three major launch facilities, as detailed in Table 2.8.

The idea to of adding the new Vostochny launch site to the two traditional cosmodromes had been in the air since the collapse of the USSR in 1991, but it received the final green light in 2007, when President Putin underlined that Russia had to ensure its "ability to carry out all kinds of space launches from [its] own territory, from automatic satellites to manned spacecraft and interplanetary probes" (The Kremlin, 2008).

The construction works started in 2011 and the new site was inaugurated in 2016. Cosmodrome Vostochny, located in the nation's Far East, is now the guarantor of Russian independent access to space and one of the biggest successes of the post-Soviet space programme. This location, which was intended to reduce Russia's dependency on the Baikonur, enables Russia to reach a wide variety of inclinations. As shown in

Table 2.9 Development phases of Vostochny Cosmodrome

Phases	Description
Phase I	Site 1S—Soyuz-2, first launch on 28 April 2016
Phase II	Site 1A (where A—Angara rocket family)—Angara-1.2/A5/A5V, launches beginning in 2021
Phase III	PU3 for a prospective super heavy-lift launch vehicle Energia-5, launches beginning after 2028

Table 2.9 following building phases have been adopted by the government vis-à-vis Vostochny's development.

2.2.3.2   Operational Satellite Programmes

The enhancement of the orbital spacecraft fleet is one of the major goals of both the FSP-2006–2015 and the FSP-2016–2025.

Earth Observation and Meteorological System

According to information provided by the Research Centre for Earth Operative Monitoring (Research Centre for Earth Observation, 2017), Russia has the following remote sensing facilities as of 2017:

- 1 Resurs-P spacecraft used for natural resources study, environmental pollution and degradation monitoring, ice conditions assessment and emergency monitoring;
- 2 Meteor-M satellites, used for Global observation of the Earth's atmosphere and underlying surface to acquire hydro-meteorological and helio-geophysical data on a regular and global basis, including agricultural and forestry problems and natural and man-made disaster monitoring;
- 2 Electro-L geostationary hydro-meteorological spacecraft designed for operational imaging of cloud cover and the Earth's underlying surface, helio-geophysical measurements, hydro-meteorological and housekeeping data collection and relay;
- 2 Kanopus V, V-IK satellites (one managed by Roscosmos and one by ROSHYDROMET/Planeta). The two satellites are designed to monitor Earth's surface, the atmosphere, ionosphere, and magnetosphere and to detect the probability of strong earthquake occurrence.

In the FSP-2025, the Russian government put special emphasis on rebuilding the remote-sensing constellation, as evident from the budget allocation of 312 billion RUB over the period. According to the new FSP, Roscosmos is planning to increase the number of spacecraft from 8 (2015) to 23 during the period to 2025. An overview is presented in Table 2.10.

According to the FSP, special attention will be paid to the exploitation of the data received from the national satellites. Consistently, in 2016 Roscosmos announced a plan to create a centre for remote sensing services that will be able to use data from all

Table 2.10 Roscosmos earth observation missions in FSP-2025

	Hydrometeorology and oceanography	Natural resources and monitoring systems
2016	Meteor M2	Resurs-P Kanopus-V
2017	Meteor M2 Arctica-M Electro-L	Kanopus V × 2
2018	Meteor-M3 Arctica-M	Resurs-P Kanopus-V × 2 Kondor-FKA Smotr-V
2019	Meteor-M3 Elektro-L4 Arctica-M2	Resurs-P Obzor-R Kondor-FKA
2020	Elektro-L5 Arctica-M2	Resurs-PM
2021	Meteor-M × 2	Resurs-PM Obzor-R
2022	Meteor-M	
2023		Resurs-PM Ozbor-R × 2
2024	Electro-L Arctica-M Meteor-MP	Resurs-PM Obzor-R
2025	Electro-M Arctica-M	Obzor-O Smotr-V

Russian remote sensing satellites (Cherebko, 2015). The centre will operate within Russian space systems' Research Center for Earth Monitoring. This company will manage all the marketing efforts and build relations with governmental and private customers. So far, the only Russian company that receives data from Roscosmos' remote sensing satellites' network is Scanex.

Communication satellites

Communication satellites have always been in the sights of the Russian space industry considering the size of the territory and the difficulty of reaching the remote localities. As of 2016, Russia had an orbital network of 32 satellites, as displayed in Table 2.11.

According to information released by the Ministry of Telecom and Mass Communications—which also participated in the elaboration of FSP 2016–2025—Roscosmos is planning to increase the number of communication satellites from 32 in 2015 to 41 by 2025. By 2030 the network should reach a total of 46 spacecraft, as displayed in Table 2.12 (Ministry of Telecom and Mass Communications of the Russian Federation, 2016).

Table 2.11 Roscosmos communications satellites as of 2016

Series	Function	Number	Orbit
Gonets-M	Telecom + Data Relay	9	LEO
Luch-5	Telecom + Data Relay	3	GSO
Ekspress	Telecom + Broadcasting	12	GSO
Yamal (Gazprom SS)	Telecom + Broadcasting	4	GSO

Table 2.12 Roscosmos telecommunication satellites in FSP-2025

	2016–2020	2021–2025	2026–2030
Satellite network	31	43	46
Active lifetime, years	10–15	15	>15

Table 2.13 Future GLONASS system

	No. of spacecraft	Series	Accuracy
2016–2020	24	GLONASS-K+K2	0.6 m
2021–2025	30	GLONASS-K+K2+VKK	0.3 m
2026–2030	30	GLONASS-K2+VKK	0.1

Navigation Satellites

Satellite navigation activities are not financed by the FSP-2025, but through a dedicated Federal Target Programme named "Development, Utilisation and Maintenance of the GLONASS System for 2012–2020" (GLONASS-2020).

The development of the GLONASS system started in the seventies with the first launches of actual satellites in 1985–1986. The system is nominally composed of 24 satellites, 21 in use and 3 spares. It was declared operational in September 1993 by President Yeltsin but the constellation was not complete until December 1995. After the completion of the system, however, Russia was not able to maintain it and the number of satellites gradually decreased down to 7 satellites in 2001. With the first Federal Target Programme GLONASS 2002–2011, the constellation reached 24 satellites in 2012, providing real-time position and velocity determination for military and civilian users with the accuracy of 4.5–7.4 m.

The current Federal Target Programme GLONASS-2020 puts special emphasis on the maintenance and exploitation (including commercial activities) of the system as well as on the development of the third generation of satellites, which can provide improved services. According to the FTP GLONASS, the budget will be decreased after the enlargement of the constellation to 30 spacecraft and the progressive replacement of GLONASS-K with GLONASS-K2 and the new GLONASS-VKK satellites, having higher accuracy and longer lifetime (see Table 2.13).

A significant part of the GLONASS programme is the ground network system. Today Russia has 14 stations located on the territory of CIS, in Antarctica, and in

Brazil. According to the system development plan, new stations will be located in Cuba, Iran, Vietnam, Spain, Indonesia, Nicaragua and Australia.

2.2.3.3 Human Spaceflight

Human space flight has traditionally remained one of the most important areas of the Russian space programme, employing at least 50% of the Federal space budget. During FSPs running from 2006 to 2015, Russia reportedly spent 186.6 billion RUB for the assembly and operation of the International Space Station (ISS). Russia's contributions and capabilities in the field are well-known: suffice it to remind the key role played by Roscosmos in maintaining the station orbit, and that all the partners of the ISS continue to rely on Soyuz for human space transportation.

Thanks to the improved funding conditions in the second half of the 2000s, the Russian government started planning new ambitious goals for its cosmonauts, including a manned lunar landing in the early 2020s and a manned Mars exploration in the late 2030s. Consistently, Roscosmos started to draft a preliminary roadmap toward the development of super-heavy launch vehicles, which closely matched the strategy followed by NASA with the Space Launch System (SLS) programme. In addition, during the 2009 Moscow Air and Space Show-MAKS, it presented the concept of a Multi-Element Interplanetary Expeditionary Complex (MEK), which would enable a manned mission to Mars, Roscosmos' top priority destination for future human spaceflight. However, by 2013 the strategy had undergone a significant change and refocused on:

- expanding the lifetime of the ISS
- building a habitable outpost at one of the Lagrangian points near the Moon.

The work of the ISS was extended until 2024 and Roscosmos came up with the plan of expanding its segment with three new modules that could be detached from the ISS and used to build a Russian Orbital Station after the decommissioning of the ISS. More specifically, the Russian segment was planned to be expanded with a Multipurpose Laboratory Module (MLM), with a launch first scheduled for 2014 but eventually postponed until March 2019 due to the clogging of the fuelling system.[25] The MLM was to be followed by the launch of the Node Module (UM), with an estimated launch in November 2019, and finally by the Science and Power Module-1 (SPM-1 or NEM-1), which is one of the two planned scientific and services modules that were constructed to become science labs and to provide the Russian segment with an independent source of power (Table 2.14).

The planned deployment of all these new and costly modules made it clear that Russia was planning to use the ISS resources as long as possible.[26] Indeed, in 2014

---

[25]Some critics, however, have claimed that the MLM project may be even abandoned after 22 years of development.

[26]This plan was supported by the call for application the new cosmonauts' corp. A second open application period was announced in 2017. It allowed every citizen of certain age and education

Table 2.14 Planned ISS modules and DSG work

Element	Launch Date	Mass	Note
Nauka—Multipurpose Laboratory Module (MLM)	2018–2019	20.3 tons	It will be docked to the ISS instead of the Pirs module and will be used for science experiments, docking and cargo. The European Robotic Arm is planned to be launched together with MLM
Pricahl—Node Module-1 (UM-1)	2019–2020	4.0 tons	A module with 6 hybrid docking ports intended to support deep space manned exploration. Can be used for the Russian OPSEK station
NEM—Science-Power Module-1 (SPM-1)	2019–2020	20.0 tons	It will be docked to Prichal and is meant to be part of the OPSEK station. It will be the most advanced component of the Russian segment including large power-generation solar arrays, laboratory facilities, living quarters, and new flight control systems
PPTS Federation	2021	16.5 tons (to the Moon)	It will be a partially reusable, piloted spacecraft for Moon exploration, orbital flights on the future Soyuz-5, instead of Soyuz and Progress
Lunar Mission Support Module (LMSM)	2025+	N.A	It will be Russian contribution to DSG for extra capabilities in life-support and storage facilities with two pressurized compartments derived from Prichal and NEM-1

Roscosmos admitted the possibility of continuing to utilise the ISS project together with NASA and the other partners up to 2030 as a test bed for future deep space exploration (e.g. for experimenting with new materials and studying human behaviour and health issues brought on by long-term living in space conditions) (Roscosmos State Corporation, 2017).

As of 2017, operations for the ISS have been extended until 2024. According to the FSP-2025, 30% of resources of the Russian segment of the ISS will be allocated to science experiments, 25% for testing new technologies and 45% for solving practical tasks such as 3D bioprinting experiments conducted by the private company 3D

level to join the current team of cosmonauts and to work at the ISS, with PTK NP and to be the first Russians who would go to the Moon.

Bioprinting Solutions and the exploitation of a multi-zone furnace for the production of large ultrapure semiconductor crystals.

The serious growth of experiments such as these can be expected in 2019 when the Russian segment is expanded with the new modules. With the restructuring of Roscosmos into a corporation, the space sector got an opportunity, and motivation, not only to conduct the research on board using the civil budget but also to make it one of the sources of income.

Parallel with ISS-related operations, Russia's human spaceflight strategy focused on the creation of technological advances for enabling exploration of the solar system in the post-ISS period. Even though the ISS was given the green light to operate until 2024, decisions about the future will have to be made well in advance, particularly if space agencies want to join forces in an effort to expand human space flight beyond the Earth orbit. In the absence of a bold commitment to go to the Moon, Mars or asteroids, space planners in Russia (and the U.S.) started to consider sending missions to the so-called Lagrange points, including a habitable outpost near the Moon which could serve as a staging hub for future deep-space exploration of the Moon or asteroids.

These plans became the basis for the discussions between Roscosmos and NASA, which eventually culminated in the signature of a joint statement on the Deep Space Gateway (DSG) at the margins of the 2017 International Astronautical Congress (IAC). Within this framework, Russia suggested developing a small airlock module for the station and assisting with transportation using the new-generation spacecraft PTK NP "Federation". The development of a new human space transportation vehicle is funded by the FSP-2025. An un-crewed version is planned to begin flight tests in 2021, while a crewed launch is scheduled to take place in 2023, from the new cosmodrome of Vostochny, aboard an Angara-5 launcher.

2.2.3.4   Exploration and Space Sciences

Space sciences and deep space exploration were key elements of the Soviet space programme, but since the 1990s they have been somehow put on the backburner as compared to other domains. Although Russian equipment has been used in a number foreign missions such as NASA's Mars Odyssey, Curiosity and LRO, the only mission led by post-Soviet Russia is the Phobos-Grunt sample return mission to one of the moons of Mars (Terekhov, 2016). The mission, launched in November 2011, ended a up in a failure, with the spacecraft re-entering the Earth's atmosphere in January 2012.

Following the unsuccessful Phobos-Grunt launch, all planetary exploration and science missions in Russia started to face delays. In January 2012, NPO Lavochkin submitted a new plan for planetary exploration to Roscosmos. As it transpired in the following months, a pair of lunar missions came to the forefront, but with launches not before 2016–2017, while other deep-space missions were pushed to the 2020s. Missions to Mars would now be limited to a possible repetition of the Phobos-Grunt project in 2018 at the earliest, and to the Russian participation in the European

Table 2.15 FSP-2025 science missions

	Budget	Name	Date	Details
Lunar exploration	38 bn RUB	Luna-Resurs-1	2019	Lunar polar orbiter
		Luna-Resurs-1 (×2)	2021	Lunar lander and rover
		Luna-Globe	2022	
		Luna-Grunt	2025	Sample return mission/search for water
Mars exploration	29 bn RUB	Exo-Mars (×2)	2016, 2020	Mars lander and rover/cooperation with ESA
		Expedition-M	2024	
Astrophysics	27 bn RUB	Spektr-RG	2017	Astrophysical labs
		Spektr-UF	2021	Astrophysical labs
		Spektr-R	–	Astrophysical labs
Heliophysics	19 bn RUB	Interheliozond		Heliosphere and sun at close range
		ARKA	2024	Stereoscopic study of the sun
		Resonans	2021	Solar corona
Biophysics	20 bn RUB	Bion-M	2021	

Dates are based on information provided by NPO Lavochkin—main developer

ExoMars launches in 2016 and 2020. The latter mission could be merged with the Mars-NET project. An asteroid-chasing mission, most likely to Apophis, remained in a definition state, with the launch date around 2020 at the earliest. Missions to Mercury and Venus were deferred to the next FTP. A very preliminary plan for the exploration of Saturn under the Saturn-TE project was also drafted at the beginning of the decade, but its preliminary development (NIR) was not expected until 2017–2019.

The plans were further revised with the approval of the FSP-2025. According to the new federal programme, Roscosmos is planning to launch 15 new spacecraft for solar system exploration. The focus continues to be on Moon and Mars missions, thus echoing the priorities set in the previous space programme (Kirkach, 2014). An overview of these missions is provided in Table 2.15.

2.3 Assessing Current Status and Future Prospects

In 2017 the Russian space sector completed a reform process intended to tackle the apparent state of systemic crisis in which it has been wallowing for the past 10 years. This reform process is certainly an indicator of the attention that the Russian government gives to national space activities. Despite decreasing funding capacities, the Russian leadership indeed continues to see space as a strategic asset that needs to be exploited for its potential.

However, it seems that the modern Russian space programme is heading in dia-metrically different directions as compared to the rest of the world. If countries such as the United States, Japan, South Korea and even India are encouraging the increasing involvement of private actors, Russia has consciously chosen not to rely on a private space sector, but rather on a public space sector, and to focus on fostering its competitiveness on international markets. The consequences are still difficult to foresee.

On the one hand, the recent centralisation may likely help in overseeing and addressing the quality issues that have arisen frequently in the past 10 years. In addition, the consolidation could have very positive impacts for the space sector by enabling a streamlining and optimisation of both infrastructure and workforce, responding to the need of improved economic performance in the light of space budget cuts, while the structural change into a state corporation could serve to ease bureaucratic requirements. On the other hand, the recent changes in the orgware have made an increasingly blurred demarcation line between policymaking, the hardware ordering and the vendor functions. This point is even more important if one considers that one of the original goals of creating URSC was to separate the client and policy aspects from the contractor sections. Without clear internal boundaries, a situation comparable to that existing before the entire reform process was started, similar to the functioning during the Soviet era, may be recreated. In addition, the change of status also means that Roscosmos is no longer obliged to publically disclose information on tenders for example, somewhat reducing the overall transparency.

In addition, even though space enterprises are supposed to become public Joint-Stock companies with shares open to investors, it remains to be seen whether the lack of de-centralisation and privatisation will support Russia's integration in global space markets (Terekhov, 2016). The current situation does not seem to bode well for the future. To illustrate this, despite having the second biggest navigation constellation, Russia's presence on the global market for navigation-related products and services is in a tiny pocket because of the lack of companies able to offer value-adding services. As a result, the socio-economic impact of public investments in GLONASS for Russia remains today limited. Indeed, whereas in the U.S., the economic activity of the navigation downstream sector is estimated to be 14 times bigger than those brought by orbital launches, in Russia such opportunities are only weakly leveraged. Similarly, Russia also has a constellation of high quality satellites for remote sensing, but their data are reportedly used for governmental purposes only. Even the biggest national map service "Yandex.Maps" still uses data from foreign satellites.

The situation is not stable even when it comes to the traditionally successful areas such as commercial launch services and scientific research. In the scientific field, for example, the number of publications related to the ISS is just 418 for Russia, compared to 2203 for the U.S., 631 for Germany, 534 for Japan and 449 for Italy). In the commercial launches field, Russia had captured a significant market share until 2013, but that share has been substantially reduced due to problems with quality control during the manufacturing phase and the related skyrocketing insurance costs as well the fiercer competition coming from the U.S. launch solutions.

All these problems bring us to the question of the efficiency of the programme itself and of private investments in the space sector in Russia, which could possibly solve a number of financing problems for Russia. The government does not seem to support private companies, though a strong public-private relationship may be essential for market development as well as for tackling many of Russia's space problems. The Skolkovo's space cluster is providing "support for private Russian companies, but numerous institutional factors in the Russian Federation will continue to challenge space entrepreneurs, and Roscosmos will likely gobble up those that show any promise. The one likely exception to this stagnation turns out to be in national security space capabilities" (McClintock, 2017). Clearly, this is the period when Roscosmos has to devise a forward-looking strategy to ensure the long-term effectiveness of its space sector.

As an indication of the importance that Russian space officials attach to the identification of effective responses to the difficult situation, on 31 March 2017, a special meeting was held to develop the strategy of Roscosmos development for the period until 2030. During this meeting, goals were prioritized, modern trends discussed, and the importance of young specialists' involvement in the industry was highlighted (Roscosmos State Corporation, 2017). At the same time, no plan of how to re-ignite the thrust of the space programme was identified. As in the past, Russia will continue to defy predictions about the collapse of its space programme, but it appears that now more than ever, it will need to further open its space programme to new actors, including international cooperation partners. The next chapters will be dedicated to the analysis of Russia's relations with the major space faring nations.

References

Al-Ekabi, C. (2015). *Space policies, issues and trends in 2014–15*. Vienna: European Space Policy Institute.

Al-Ekabi, C. (2016). *The future of European commercial spacecraft manufacturing*. Vienna: European Space policy Institute.

Aliberti, M., & Tugnoli, M. (2016). *European launchers between commerce and geopolitics*. Vienna: European Space Policy Institute.

Astrowatch. (2014, March 4). *Russian Prime Minister inks asset merger for New Space Corporation*. Retrieved October 2, 2017 from Astrowatch: http://www.astrowatch.net/2014/02/russian-prime-minister-inks-asset.html.

Bloomberg. (2017, December). *RUB to EUR exchange rate*. Retrieved December 1, 2017 from Bloomberg markets: https://www.bloomberg.com/quote/EURRUB:CUR.

Bodner, M. (2015, January 23). *Putin approves Roscosmos merger with conglomerate*. Retrieved October 8, 2017 from Space News: http://spacenews.com/putin-approves-roscosmos-merger-with-conglomerate/.

Bodner, M. (2017, October 4). *60 years after Sputnik, Russia is lost in space*. Retrieved October 5, 2017 from Space News: http://spacenews.com/60-years-after-sputnik-russia-is-lost-in-space/.

Brjanski, G. (2017, September 7). *Igor Komarov: We are planning to start the launches from Vostochny in 2019*. Retrieved November 18, 2017 from TASS: http://tass.ru/vef-2017/articles/4539362.

Cherebko, I. (2015, June 18). *Russia will create a global company for remote sensing*. Retrieved November 8, 2017 from Izvestia: https://iz.ru/news/587864.

Cherebko, I. (2016, January 13). *Restrictions on Roscosmos employees for leaving the country*. Retrieved November 10, 2017 from Izvestia: https://iz.ru/news/601394. Accessed February 14, 2018.

Clark, F. (2013). Reforming the Russian Academy of Sciences. *The Lancet, 382,* 1392–1393.

Homchenko, J. (2017, April 12). Space Issues. *BRC + Space Industry*.

Interfax. (2017, June 1). *Roscosmos will launch the first Soyuz-5 rocket in 2022*. Retrieved Decemer 1, 2017 from Interfax: http://www.interfax.ru/russia/564765.

International Astronautical Federation. (2014). *United Rocket and Space Corporation*. Retrieved October 15, 2017 from International Astronautical Federation: http://www.iafastro.org/societes/united-rocket-and-space-corporation/.

IZ Russia. (2016, June 17). *Roscosmos will agree on the development strategy by the end of the year*. Retrieved Novemer 10, 2017 from IZ Russia: https://iz.ru/news/618516.

Kirkach, S. (2014, August 2). *Lev Zeleny: Space researchers will concentrate on moon and mars*. Retrieved November 19, 2017 from RIA News: https://ria.ru/science/20140802/1018573356.html.

Kosenkov, I. (2016). Role of Skolkovo in the development of Russian private space industry. In *Symposium looking to the future: Changing international relations and legal issues facing space activities*. Vienna: European Centre for Space Law.

Levada Center. (2017, December). *Putin's approval rating*. Retrieved December 23, 2017 from Yuri Levada Analytical Center: https://www.levada.ru/en/ratings/.

McClintock, B. (2017). The Russian space sector: Adaptation, retrenchment, and stagnation. *Space and Defence, 10*(1), 3–8.

McGilvray, A. (2016, May 6). *World's oldest science network faces uneasy future*. Retrieved January 31, 2018 from Nature Index: https://www.natureindex.com/news-blog/worlds-oldest-science-network-faces-uneasy-future.

Ministry of Telecom and Mass Communications of the Russian Federation. (2016, March 18). *Official statement: We are planning to enlarge the constellation of communication satellites up to 41 by 2025*. Retrieved November 9, 2017 from Minsvyav: http://minsvyaz.ru/ru/events/34852/.

Moskowitz, C. (2011, August 30). *Report: Russia identifies cause of rocket launch failure*. Retrieved October 7, 2017 from Space.com: https://www.space.com/12779-russian-rocket-failure.html.

Research Centre for Earth Observation. (2017). *List of remote sensing facilities in Russia*. Retrieved November 18, 2017 from Research Centre for Earth Observation: http://eng.ntsomz.ru/ks_dzz/satellites.

Reuters. (2013, December 26). *Russia bets on sweeping reform to revive ailing space industry*. Retrieved October 7, 2017 from Reuters: http://www.reuters.com/article/2013/12/26/us-russia-space-idUSBRE9BP02S20131226.

RIANOVOSTI. (2013, July 4). *Russia's space program is ineffective—Audit chamber*. Retrieved October 1, 2017 from RIANOVOSTI: http://en.ria.ru/russia/20130704/182063035.html.

Roscosmos State Corporation. (2017, March 31). *Strategic development of the State Corporation ROSCOSMOS for the period before the year 2025 and for the perspective up to the year 2030,*

Moscow, 31 March 2017. Retrieved November 30, 2017, from Roscosmos State Corporation: https://www.roscosmos.ru/media/files/docs/2017/doklad_strategia.pdf.

RSC Energia. (2018). *S.P. Korolev Rocket and Space Corporation Energia.* Retrieved January 13, 2018 from RSC Energia: https://www.energia.ru/en/corporation/oao.html.

Space News. (2013, December 9). *Putin signs legal decree consolidating Russian industry.* Retrieved October 6, 2017 from Space News: http://www.spacenews.com/article/civil-space/38551putin-signs-legal-decree-consolidating-russian-industry.

Stanovaya, T. (2013, July 15). *Reform of the Russian Academy of Sciences: Checkmate in two moves.* Retrieved January 31, 2018 from Institute of Modern Russia: https://imrussia.org/en/nation/513-reform-of-the-russian-academy-of-sciences-checkmate-in-two-moves.

TASS News Agency. (2013, December 3). *Overhaul pending in Russian space sector.* Retrieved October 8, 2017 from ITAR-TASS: http://en.itar-tass.com/russia/763105.

TASS News Agency. (2017, May 22). *Source reveals: The first launch of the Federation spacecraft is planned to be postponed till 2022.* Retrieved November 15, 2017 from TASS: http://tass.ru/kosmos/4287423.

Terekhov, P. (2016, December 19). *Russia is losing its place in the growing space economy. Or not?* Retrieved November 12, 2017 from GeekTimes: https://geektimes.ru/post/283810/.

The Kremlin. (2008, April 11). *Opening remarks at a meeting with the security council on Russia's space exploration policy for the period through to 2020 and beyond.* Retrieved November 11, 2017 from The Kremlin: http://en.kremlin.ru/events/president/transcripts/24913.

The Moscow Times. (2017, January 25). *Russian police investigate alleged substitution scam at Rocket Engine Factory.* Retrieved October 4, 2017 from The Moscow Times: https://themoscowtimes.com/news/experts-check-russian-rocket-engines-for-low-quality-metal-56918.

The World Bank. (2017, December). *GDP-Russian Federation.* Retrieved December 8, 2017 from The World Bank: https://data.worldbank.org/indicator/NY.GDP.MKTP.CD?locations=RU.

Toporov, A. (2017, Feruary 3). *Manual labour's fault in the rocket crashes 3 February 2017.* Retrieved November 1, 2017 from Vzglyad: https://vz.ru/society/2017/2/3/856514.html.

Zak, A. (2012, November 26). *The Russian space industry at the turn of the 21 century.* Retrieved October 7, 2017 from Russian Space Web: http://www.russianspaceweb.com/centers_industry_2000s.html.

Zak, A. (2015). *Angara-5 to replace proton.* Retrieved October 10, 2017 from Russian Space Web: http://www.russianspaceweb.com/angara5.html. Accessed December 11, 2015.

Zak, A. (2017a, July 21). *Angara -5P launch vehicle.* Retrieved November 1, 2017 from Russianspaceweb: http://www.russianspaceweb.com/angara5p.html.

Zak, A. (2017b, July 24). *Super-launcher is back on the books.* Retrieved November 13, 2017 from Russianspaceweb: http://www.russianspaceweb.com/energia5v.html.

Zhukov, S. (2015, January 23). *Роскосмос умер. Да здравствует «Роскосмос (Roscosmos is dead. Long live Roscosmos).* Retrieved November 13, 2017 from Zelenyikot: http://zelenyikot.com/roscosmos-dead/.

Zyatkov, N. (2014, December 23). *Дмитрий Рогозин: вооружаем армию, чтобы не воевать (Dmitry Rogozin: arm the army, not to fight).* Retrieved November 13, 2017 from AIF: http://www.aif.ru/politics/russia/1413030.

Chapter 3
The External Evolution of the Russian Space Programme

3.1 Russia's International Posture: Drivers and Evolution

From the Intercosmos programme to the currently envisaged Lunar Orbital Platform-Gateway (LOP-G), passing through the International Space Station (ISS) and the 2016 ExoMars mission, Russia has traditionally positioned international space cooperation as an integral component of its space programme and strategy. Russia's contribution to the space endeavours of India, South Korea, Brazil, the U.S., Europe and China "bears witness to Russia's position as a linchpin of most current international space architecture" (Hulsroj, 2014). In spite—or perhaps exactly because—of the tumultuous vicissitudes experienced by its domestic space sector recently, Russia has always remained open to international cooperation and has repeatedly demonstrated the importance it gives to meeting its international commitments, even during difficult times, as shown *inter alia* by the example of the ISS (Mathieu, 2008).

3.1.1 Russia's Drivers for Space Cooperation

Russia's policies in the international space arena have generally been driven by a combination of programmatic drivers and broader policy goals including funding needs, technological needs, industrial and commercial purposes, as well as foreign policy and strategic objectives.

From a programmatic perspective, most of Russia's recent space cooperation undertakings have responded to the logic of coping with funding requirements. As happened in the 1990s, Russia's ability to fund its space activities has sensibly reduced over the past five years, but its plans have nonetheless remained rather ambitious. Therefore, cooperation is seen as an instrumental source of funding to achieve ambitious objectives under funding cuts, particularly in areas neglected by the current FSP2025, such as space sciences and robotic exploration (e.g. cooperation with the European Space Agency on the Luna 26 programme).

© Springer International Publishing AG, part of Springer Nature 2019
M. Aliberti and K. Lisitsyna, *Russia's Posture in Space*, Studies in Space Policy 18,
https://doi.org/10.1007/978-3-319-90554-9_3

While cooperation enables partners to share/reduce the cost burden of projects they might otherwise do on their own, in Russia's case, considering the recent vicissitudes of its space sector, a key element is that cooperation makes its programmes more stable and ensures national funding and a certain degree of continuity in activities (Mathieu, 2008). International cooperation has been more broadly identified as a way of both aggregating resource for larger projects and increasing the effectiveness of its national programme, by freeing up resources and allowing Roscosmos to match its resources more effectively to its plans.

Moscow has also been particularly eager to use its leading position and expertise to derive material benefits from its partnerships, thereby generating alternative revenues for financing national space activities. Technology transfers and supply of its products to foreign partners have in part responded to this logic. Although in the course of the last decade Russia opted to move away from the "buyer-seller" type of relations adopted in the 1990s, recent difficulties in financing space activities have urged it to reconsider this practice.

Alongside funding-related needs, cooperation has been a way for Russia to broaden its sources of know-how and expertise. Scientists and engineers from Western countries in particular provide Russian teams with access not only to technologies, but also to know-how vis-à-vis industrial processes that could improve the performance of Russia's space systems and allow the country to gain/develop state-of-the-art technologies that it lacked by further cooperating with these foreign partners. Notably, Roscosmos has expressed its interest in joining the European Cooperation for Space Standardisation (ECSS).

However, international cooperation has not only been leveraged for the pursuit of solely programmatic goals. By both design and opportunity, Russia has been very successful in linking space cooperation to the pursuit of broader policy goals.

In the eternally perceived need to "catch-up" to the economic achievements of its counterparts, cooperation has been used to serve the objectives of technology-led macro-economic development and modernisation. International partnerships, with the West in particular, have been seen as a way to gain know-how and to stimulate the overall development of the industrial base, but also as a way to maximise the benefits of investments in the space industry.

Most visibly, however, space cooperation has been used as a tool to fulfil political and strategic purposes in the country's foreign policy and standing in the international arena. As contended by Andrew Kuchins and Igor Zevelev, despite Russia's international status having experienced particularly wide swings in the past 40 years, "certain elements of Russian national identity and core foreign policy goals find their roots and continuity deep in Russian history. And one of them is the enduring belief that Russia is a Great Power and must be treated as such" (Kuchins & Zevelev, 2012). Clearly, such eagerness to affirm (or reaffirm) its great power status in its foreign policy can also be seen in the space field. Participating in large cooperative undertakings (e.g. the ISS) has been a way for Russia to generate diplomatic prestige Consistently, the Russian leadership has typically sought international partnerships to reflect Russia's national interests and priorities, thereby requesting more control

Table 3.1 Drivers of Russia's space cooperation

Driver		Objective
Programmatic drivers	Funding needs	Provide funding for joint activities or revenues for financing the FSP (e.g. RD-180)
	Technological needs	Improve the performance of its systems and access state-of-the-art technologies (e.g. ISS)
Broader policy drivers	Domestic policy objectives	Increase the stability and continuity of national space activities (e.g. Mars exploration)
	Foreign policy objectives	Strengthen political ties and increase influence over partner (e.g. launch of Iran's satellites)

over the terms of their cooperation as well as a real say in the decisions, if not actually having leadership.

The Russian leadership has also visibly used space as a foreign policy tool, especially to strengthen political ties with selected countries or to affirm the country's regained power and influence over those countries. Russia has indeed been at the forefront of leveraging its expertise in space technology (particularly launch and human spaceflight technology) as a means to extend its political influence (consider for instance its cooperation with Brazil, India or China). Last, but not least, Russia has been diversifying its international partnerships as a way of strengthening broader political ties and President Putin's move towards multi-vectoral foreign policy has been duly reflected in its space cooperation as its list of partners in the field since the mid-2000s has started to grow longer (Mathieu, 2008).

Table 3.1 summarises the drivers of Russia's cooperation with foreign partners, though Russia's various partnerships are driven by a combination of these rationales rather than a single specific objective. To illustrate, the numerous technology transfers provided by Russia to a variety of emerging spacefaring nations have been a tool to both increase revenue for funding its space programme and also a means to exercise influence over the partner.

3.1.2 Types of Space Cooperation

The role of Roscosmos has been essential in the development of international cooperation, especially given that, in Russia, international cooperation generally takes place within the framework of intergovernmental agreements and is always overseen by

Table 3.2 Relevant technology transfers from Russia in the field of launch vehicles

Country of destination	Manufacturer	Component	Purpose
U.S.	Energomash	RD-180 engine	Atlas V
U.S.	Energomash	RD-181 engine	Antares
U.S.	Kuznetov DB	NK-33 engine	Antares
India	Glavkosmos	C-12 engine	GSLV Mk-I
North Korea	Korolyev OKB	Scud Missile plans	No-Dong
South Korea	Khrunichev	Angara technology	KSLV-I
Ukraine	Energomash	RD-251/2	Cyclone
Brazil	Energomash	Unspecified upper stage	VLS-1

the Federal Space Agency. The Federal Space Agency also itself implements some cooperation activities (Mathieu, 2008). This is even more the case today, as former separate elements of the programme have been centralised within Roscosmos SC.

Over the years, Roscosmos has engaged foreign partners through different cooperation schemes with a variety of objectives ranging from data/information exchange and policy dialogues to joint programmes, passing through provision of space-related services and technology transfer. In this respect, it is important to stress that Russia does not have the same definition as Europe of what cooperation is. Indeed, whereas Europeans generally consider cooperation as being activities based on a non-exchange of funds (barter agreements), Russians tend to include as 'cooperation' sales of technologies and products. As detailed below, the main types of cooperation undertaken by Russia are:

• Technology transfer
• Joint ventures
• Joint programmes
• Exchange of data
• Policy dialogue.

Russia has been at the forefront in advancing bilateral ties with selected countries and increasing revenue for its programmes by means of technology transfers. For example, it has transferred entire plans of its Scud missile to the Democratic People's Republic of Korea; provided—amid a strong U.S. opposition—cryogenic engines to India, and an unspecified liquid upper stage to Brazil; designed South Korea's KSLV-1 first stage; and marketed several engine technologies to the U.S. (see Table 3.2 for an overview of the major transfers in the field of launch vehicles).

International cooperative undertakings have also taken the form of joint ventures for the commercial exploitation of Russian launch vehicles. Such undertakings have responded to both commercial and political drivers. The first international joint ventures were created in the mid-1990s with the aim of exploiting launch vehicles developed in the former USSR. Their creation proved to be a good compromise between

Table 3.3 Russia's international joint ventures for launch services

Name	Year	Original composition	Service product
Sea Launch	1995	Boeing (U.S.) RSC-Energia, (Russia) Aker Solutions ASA, Yuzhnoye SDO (Ukraine) Yuzhmash PO (Ukraine)	Zenit-3SL Zenit-3M
International Launch Services (ILS)	1995	Lockheed Martin (U.S.) Khrunichev (Russia) RSC-Energia (Russia)	Proton, Angara
Space International Services (SIS)	1996	RSC Energia (Russia) UGMK (Russia) Yuzhnoye SDO (Ukraine) Yuzhmash PO (Ukraine) TsENKI (Russia)	Zenit-2SLB Zenit-2SLB Zenit-3F
International Space Company (ISC)	1997	Roscosmos (Russia) Ukranian Space Agency (Ukraine) Garysh Sapary (Kazakhstan)	Dnepr-1
Starsem	1996	EADS Astrium (Germany) Roscosmos (Russia) TsSKB-Progress (Russia) Arianespace (France)	Soyuz
Eurocktot Launch Services	1995	EADS Astrium (Germany) Khrunichev (Russia)	Rockot

divergent interests, as they allowed Western countries to maintain some indirect control over new competitors in the launch market, while enabling the Russian partner with insufficient national funds to secure finance (to be re-injected into national space activities) and acquire hands-on commercialization know-how. The four major joint ventures, namely International Launch Services (ILS), Eurockot Launch Services, Sea Launch and Starsem, were set up between 1995 and 1997. In addition to these, a joint venture within the former Soviet Union, the International Space Company (ISC) Kosmotras, was created in 1997 by Russian, Ukrainian and Kazakh entities in order to ensure continuity of launch operations (see Table 3.3). An overview of the evolution of these international joint ventures is provided in Annex C.

Over the past years, cooperation activities have been undertaken by Russia mostly in the form of **joint programmes** or participation in foreign missions. The most relevant examples include, of course, the ISS, but also extend to notable programmes such as the joint ExoMars missions with the European Space Agency (ESA), the Phobos-Grunt mission with China, and the participation in NASA's Mars Science Laboratory. Interestingly, joint cooperation initiatives have also covered the development/exploitation of ground infrastructure. The utilization of launch sites on foreign territory has been of particular relevance for Russia, which has signed an agreement with Kazakhstan to ensure a long-term lease of the Baikonur Cosmodrome and its associated facilities until 2050, at an annual fee of $115 million. Particularly noteworthy is also the exploitation of Soyuz-ST from the Guyana Space Centre (GSC). As already highlighted in a previous ESPI study, "such cooperative undertaking was of interest to both Europe and Russia, as it enabled the former to complement the performance of ESA launchers Ariane and Vega, and the latter to benefit from improved access to commercial markets and improved performance of the Soyuz launcher when being launched much closer to Equator. At that time, it was also in the interest of both Russia and Europe to boost their relationship, and cooperation in the launcher sector was seen as a tool for pursuing this objective" (Aliberti & Tugnoli, 2016).

Other forms of international cooperation featuring Russian involvement concern the exchange of data and information. For instance, Russia has been sharing data from its meteorological satellites with the European Organisation for the Exploitation of Meteorological Satellites (EUMETSAT) and has also been providing EO data within the framework of the International Charter on Space and Major disasters. Similarly, Russia's GLONASS system has contributed to the COSPAS-SARSAT system, while in the field of SSA, it has set up the International Scientific Optical Network (ISON), an international project currently consisting of about 30 telescopes and about 20 observatories in about 10 countries which have been organized to detect, monitor and track objects in space (Molotov, Voropaev, & Borovin, 2016).

Finally, mention must be made of cooperative exchanges in terms of policy dialogue and diplomatic initiatives. For instance, following the 2009 satellite collision between a defunct Russian satellite and a U.S. commercial telecommunications satellite, U.S. and Russian space policy officials and experts have started to engage in bilateral expert meetings to discuss issues of transparency and confidence-building measures (TCBMs) in outer space. From a relatively different standpoint, Moscow and Beijing's authorities have engaged in space policy and security dialogues, which culminated to the submission, in 2008 and 2014, of a "Draft Treaty on the Prevention of the Placement of Weapons in Outer Space", to the Conference on Disarmament (CD), a joint diplomatic initiative primarily aimed at contrasting the hegemonic approach of the Bush Administration to space matters.

3.1.3   Russia's Evolving Attitude to Cooperation

The recent evolution of the Russian space sector and the recent strains in Russia's economic performance and international relations (e.g. Ukraine, Crimea, Syria) have led to an evolution of its cooperation with foreign partners. Importantly, these developments have not directly impacted the foreign policy and programmatic space drivers pursued by Russia, but have rather altered Moscow's posture to achieving those objectives. From an overall perspective, the evolution of Russia's international posture can be seen as having simultaneously gone in two diverging directions:

- On the one hand, there has been a sensible reduction of cooperation with several former partners (most visibly, Ukraine, Kazakhstan and the U.S.) accompanied by striving towards greater autonomy from foreign sources. This tendency intensified following the 2014 Crimea crisis, a year that marked the end of the cooperative undertakings that Russia put in place in the mid-1990s. More specifically, Russian space authorities:

  - repatriated the share of the multinational joint ventures to Russian entities,
  - put a stop to future use of vehicles not entirely manufactured domestically, like Dnepr and Zenit (also in light of the prior prohibition on the Ukrainian side to export to Russia missile and launcher components)
  - started to develop production capabilities for critical components that were not yet domestically available.

- On the other hand, there has been a broadening of cooperation with new actors, particularly emerging spacefaring nations such as Brazil and South Korea, and non-Western countries such as China, India and Iran, in the form of both technology transfer and joint activities. Such diversification in Russia's partnership portfolio was admittedly already unfolding in the mid-2000s but has further intensified in recent years, in connection with Russia's worsening economic performance and diplomatic relations with the West. The partnership diversification has been driven by different factors, namely by:

  - Russia's resolve to counterbalance the increasing strains on cooperation with the West, and indeed to mitigate the risks of isolation in Russian international outreach in space. For instance, Russia quickly turned to China as a back-up partner in the area, including provision of electronic components.
  - The intention to fill the gaps in the cooperation avenues left internationally open by West, and hence as a way of strengthening political ties with non-Western powers and groupings (e.g. the BRICS).
  - The increasing need to find alternative sources of revenue because of the budget reduction of recent years. Emerging space faring nations such as Brazil, Iran and Korea want to develop their space activities quickly and need cooperation to acquire technologies and expertise. Russia transfers technologies and know-how to these partners and does joint work. However, cooperation between Russia and these countries is considered as potentially sensitive in some cases, both domestically and internationally, and issues of technological transfer tend to hinder it.

The complex evolution of Russia's relations with established and key emerging players will be presented in the following section.

3.2   Evolution of Russia's Space Relations with Key Players

After highlighting the general drivers and unfolding trends of Russia's international space posture, this second section provides a more specific account of the evolution and current standing of the country's relations with the international space community. Particular attention is devoted to the U.S., China and India—having been (together with Europe) key partners for Russia—but minor cooperation (or competition) axes are also assessed.[1] Similarly, the assessment covers overall historical developments in Russia's bilateral space diplomacy since the end of the Cold War, but puts the spotlight on its most recent evolution until the end of 2017.

3.2.1   Russia-U.S. Space Relations—Bound to Cooperate?

Russia's relationship with the United States has arguably been the most important bilateral relationship in the space arena, a relationship that, over the past 60 years, has set the trends and shaped the priorities and timing of the international space agenda. After all, both the space race and the cooperation paradigm that emerged in the 1990s are distinctive by-products of Russia-U.S. interaction. Such interaction has always been complex and closely intertwined with wider political relations. But even though the two countries are now living through what has been described as a new Cold War, urging both Moscow and Washington to reconsider the scope of their space ties, cooperation has not come to an end, and actually may be further solidified in some space activity segments in the future.

An account of the evolution of Russia-U.S. space relations seems indeed to suggest that no matter how strained the political interplay may have turned in recent years, the relations of mutual dependence (and hence mutual vulnerability or mutual interest) that the two partners have reached over the years make them bound to cooperate.

3.2.1.1   Russia-U.S. Space Cooperation Overview

From an overall perspective, Russia and the U.S. have promoted selective—yet ground-breaking—cooperation in the field of space activities over the 20 years since the collapse of the Soviet Union. The cornerstone of such cooperation has been, of course, the International Space Station (ISS). Although the core activities were undertaken during the 1990s and early 2000s, Russo-American collaboration once again became crucial following the decommissioning of the Space Shuttle fleet in 2011, when the United States was left with no domestic ability to provide crew access to the ISS.

While financing the commercial development of a follow-on crew space transportation system that could provide a safe, reliable, and cost-effective transportation

---

[1] Specific considerations on Europe-Russia relations will be provided in Chap. 4.

Table 3.4 Total cost of Soyuz seat per year

Launch year	Number of seats	Total cost ($ million)
2006	2	$50.200
2007	1	$21.800
2008	1	$21.800
2009	6	$150.997
2010	6	$158.550
2011	6	$224.426
2012	6	$306.000
2013	6	$335.070
2014	6	$361.875
2015	6	$364.869
2016	6	$424.045
2017	5	$381.641
2018	7	$567.500
Total	64	$3368.774

Source NASA Office of Inspector General, (2017)

to and from the ISS, the National Aeronautics and Space Administration (NASA) has since then relied on purchasing seats from Roscosmos on its Soyuz spacecraft to maintain a U.S. presence on the station.

Logically, Russia has shrewdly exploited this dependency situation of NASA, by substantially raising the price it charges for the Soyuz services. As reported by the NASA Office of Audits (NASA Office of Inspector General, 2017), the round-trip cost for a seat on the Soyuz has increased approximately 384% over the last decade from $21.3 million in 2006 to $81.9 million under the contract modification signed in August 2015 (see Table 3.4).[2]

Russian and American interaction on the ISS programme, however, has not been limited to ferrying astronauts to the station. Over the past 15 years, scientific collaborations have remained constant,[3] and representatives of both countries' space agencies are present in Mission Control Moscow and Houston at all times to support and the ensure the safety of each other's operations on the ISS.

In addition to the ISS, the Russian scientific community has been cooperating with NASA also on robotic exploration of the Moon and Mars, most recently on NASA's Lunar Reconnaissance Orbiter (LRO) and on NASA's Mars Science Laboratory (MSL), which were respectively launched in 2009 and 2012 and are still in operation today. The Space Research Institute (SRI) of the Russian Academy of Sciences (RAS) provided instruments on both the spacecraft, including IKI's Dynamic

---

[2]This is in part also due to the fact that the Russian rouble has undergone significant fluctuations over the past decade.

[3]For instance, Scott Kelly and Mikhail Kornienko spent almost a full year in space to study the effects of long duration flight.

Albedo of Neutrons (DAN) instrument, which is searching for water on the Curiosity Rover, and the Lunar Exploration Neutron Detector (LEND) instrument which is searching for water from NASA's LRO in lunar orbit.

Mention must be made also of another and more ambitious programme that has been under consideration over the past five years: Russia's Venera-Dolgozhivuschaya (Venera-D). As explained by Sergey Oznobishchev, "the Venera-D project originated from the desire to repeat and multiply the successes of the Soviet Union's domestic programs in the 1970s to explore other planets. It was initially assumed that the project to explore Venus, run by the SRI, together with a number of other renowned Russian institutions and the support of the Lavochkin Research and Production Association, would be implemented over the next decade on the basis of broad international cooperation", involving both European and American partners (Oznobishchev, 2014).

Another long-standing field of cooperation between Russia and the U.S. has been access to space. Such cooperation has taken the form both of exploitation agreements of Russian rockets on the commercial launch markets and the supply or Russian rocket engines to the U.S. An overview of the International ventures featuring Russian participation has been already presented in previous section (see Sect. 3.2). As for the transfer of Russian engine technology for use on Atlas V first stage and strap-on boosters, this cooperation has been managed on the basis of a Joint Venture (JV) between the engine developer Energomash and Pratt & Whitney Rocketdyne.[4] This 50-50 JV company, RD AMROSS LLC, was created in 1997 for marketing, sale, shipment and support services for RD-180 engines and the organisation of a U.S. industrial base for parallel production of RD-180 engines and their modifications. To adapt the Atlas V requirements, the original four-chamber Energia RD-170 was modified into a two-chamber engine named RD-180.

According to the RD-180 Engineering and Manufacturing Development contract awarded in 1998, Pratt & Whitney was to act as funding source for RD-180 development, representing the U.S. Co-Production source, and Energomash assumed the development and design role, producing RD-180 engines for Lockheed Martin. The first serial production RD-180 engine was delivered to the USA in 1999 while U.S. production was to be enabled by 2002.

However, following a transfer of detailed engineering and manufacturing technical documentation from NPO to P&W Rocketdyne in 2004, no further steps to establish U.S. production capacity have been taken due to: (a) high costs that would have been associated with the establishment of such a capacity; and (b) the fact that the details provided by the Russians were in the final analysis considered insufficient to build the engine in the U.S. Consequently, after coordination with the U.S. Air Forces (USAF), the RD-180 engine co-production programme was concluded in 2008, in part as a reaction to the 2008 conflict in Georgia. While the USAF recommended

---

[4]Pursuant to the DoD policy of 1995 on the use of Former USSR's propulsion in orbital launch vehicles according to which "FSU [Former Soviet Union]-produced propulsion systems [...] used in launch vehicles for national security missions must be converted to U.S. production within four years after contract award for Engineering and Manufacturing Development", a 1997 Russian presidential decree authorised Energomash sales of RD-180 engines in the U.S. and the organisation of parallel manufacturing in the USA.

to the DoD the continued use of the RD-180 engine on the Atlas V without a full U.S. production facility, to mitigate the risk of reliance on foreign suppliers it was decided to stockpile the engines.[5]

3.2.1.2 Announcing Divorce, Preparing to Cooperate

By the end of the 2000s, this and other cooperation activities turned increasingly lukewarm, in part also as a result of the increasingly difficult relations between the Medvedev-Putin tandem and the Obama administration.

In the field of access to space, international joint ventures were restricted to the marketing of already operational Russian launch systems, with no involvement of the Westerns partners in either the launcher system or the associated launch complexes. Further, Russia regained total control of commercial exploitation by repatriating the share of Sea Launch and ILS ownerships. As for U.S.-Russian cooperation on Atlas V rocket engines, in 2011, the Russian Comptroller Office faulted Energomash for having sold RD-180 engines at half their production cost resulting in a loss of US$32 million, which represented 68% of Energomash's total losses. In April 2013, the exclusivity of the RD-180 arrangement was questioned as it was seen to prevent open-market sales of RD-180 engines.

The move was apparently also linked to the fact that in 2013 the Americans refused to host GLONASS ground stations in U.S. territory due to national security concerns (Gibbons, 2014; Miller, 2014). They also imposed a ban on the purchase of electronic components to Russia in connection with the scandal involving former NSA employee Edward Snowden (Oznobishchev, 2014).

Relations further deteriorated in 2014, following Russia's purported backing of the armed insurgency in eastern/southern regions of Ukraine that resulted in the eventual annexation of the Crimean Peninsula. On 2 April 2014, NASA announced the suspension of all contacts with Russian entities and of all cooperation activities, with the sole exemption of ISS-related activities. The ban barred NASA employees from travelling to Russia, hosting visitors and even e-mailing or holding teleconference with the Russian counterparts. The impact of this ban mainly concerned scientific exchanges with RAS on NASA's MSL, the on-going discussion on the potential cooperation on the Venera-D programme, the provision of electronic components to Russia, as well as the initiated discussions towards the interoperability of GPS and GLONASS. In addition, while the exemption of ISS cooperation underscored the dependence of the U.S. on Russia, NASA made it clear that it was "laser focused on

---

[5]Under a five-sided Agreement between the Russian space agency, NPO Energomash, Lockheed Martin, RD AMROSS and Pratt & Whitney of May 1997, Lockheed Martin committed to purchasing 101 RD-180 engines. The currently valid exclusive fixed-price 5-year block buy contract foresees delivery of 5–6 engines per year through 2018. By Oct. 2012, 59 engines had been delivered and Energomash planned to supply another 29 RD-180 engines between 2014 and 2017, foreseeing a production rate of 4–5 engines per year. Energomash also stated that it was "now delivering the engines at a price that is three times higher than the 2009 price". By end May 2014, the stock had reached a level of 16 engines with orders for 11 additional engines having been placed.

a plan to return human spaceflight launches to American soil, and end our reliance on Russia to get into space" (National Aeronautics and Space Administration, 2014).

While immediately extending cooperation with Chinese partner in many of these areas (see next section), as a retaliation Russian Deputy Prime Minister Dmitry Rogozin soon after announced that "Russia would disable the ground stations of America's GPS system situated within its territory, halt supplies to the U.S. of RD-180 rocket engines if they continue to benefit the Pentagon, and oppose keeping the ISS in operation after 2020" (Luzin, 2014). Even though NASA purportedly had no contingency plan in case Russia decided to stop ferrying American astronauts to the station, that option was not raised by Rogozin, clearly aware of the political blowback this would generate and of the impossibility of Roscosmos continuing to operate the station without NASA.[6] More broadly, it can be noted that Moscow was not interested in putting forward these steps, as they would have eventually damaged Russia much more than the Western partners. In addition, Russia was—and still is—motivated by the will to appear a solid and reliable partner, irrespective of political misalignments on the international stage.

However, primarily because of the U.S. striving to eliminate its dependence on Russia with respect to both the ISS and its Atlas V launch vehicles, cooperation activities seemed well on track to coming to an end. By 2015 NASA indicated it had no plans to review the manned transportation contract signed with Roscosmos that was set to expire in 2018 (Musa, 2016). Similarly, the RD-170/180 cooperation was officially halted through provisions in the U.S. National Defense Authorization Act of 2015, which barred U.S. launch providers contracting with Russian engine suppliers for the U.S. Expendable Launch Vehicle programme (EELV)—(with the exception of orders already placed under the 2013 EELV block buy contracts).

Although public funds were immediately made available for U.S. engine development (and alternative U.S. engines were identified), by 2016 a 360° course reversal started to emerge. The RD-170/180 cooperation was in fact, re-enabled again through the 2016 Appropriations Act. This re-authorisation was mainly motivated by the difficulties a lack of engines would pose for the exploitation of Atlas V and therefore for the continued availability of competing launch solutions for U.S. institutional payloads. Subsequently, ULA placed an order for 20 additional RD-180 engines with Energomash in early 2016, which were to enable another 24 DoD launches and 25 civil institutional or commercial launches, taking into account already existing engine stock. A summary of the of RD-180 cooperation's evolution is provided in Fig. 3.1

A similar reverse of course in Russia-U.S. space relations occurred with respect to robotic space exploration missions and to the ISS-related activities. At the end of 2016, it was in fact disclosed that discussions between RSA and NASA for the Venera-D programme had been re-activated and that a joint team of Russian

---

[6] As also explained by Bolden before members of Congress during a hearing on 27 March 2014. "The partners would probably have to shut the space station down. If you're thinking that the Russians will continue to operate the International Space Station, it can't be done [...] because we provide navigation, communications, power..." (Kramer, 2014).

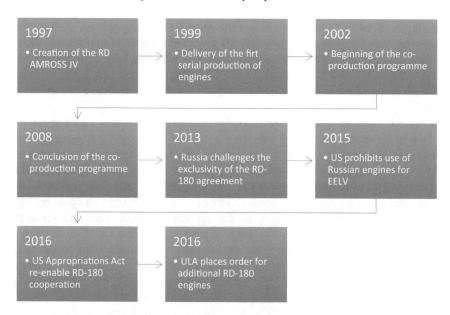

Fig. 3.1 Evolution of US-Russia cooperation on RD-180 engines

and American scientists were assessing and refining the scientific objectives of the mission to Venus, currently planned for 2025. In January 2017, the joint science-definition team issued a report on the programme (Senske & Zasova, 2017) and in October 2017 a modelling workshop was held at the Russian Space Research Institute to contribute to decisions about the mission's scientific instruments.[7]

As for the ISS, in 2016 the Russian government officially accepted the NASA-led proposal to extend the life of the ISS until 2024. On the U.S. side, the delays of Boeing and SpaceX in the development of their crew transportation systems, eventually left NASA with no options other than contracting a new agreement with Roscosmos for ferrying its astronauts to the station. The agreement was contracted through Boeing as part of the modification to the Vehicle Sustaining Engineering Contract that NASA has with Boeing for the ISS operations (Foust, 2017a, b). The agreement covered two Soyuz seats on flights in the fall of 2017 and spring of 2018, with an option for three seats on Soyuz flights in 2019.[8] It is also important to stress that even

---

[7] As reported by NASA, "the IKI Venera-D mission concept as it stands today would include a Venus orbiter that would operate for up to three years, and a lander designed to survive the incredibly harsh conditions a spacecraft would encounter on Venus's surface for a few hours. The science definition team is also assessing the potential of flying a solar-powered airship in Venus's upper atmosphere. The independent flying vehicle could be released from the Venera-D lander, enter the atmosphere, and independently explore Venus's atmosphere for up to three months (National Aeronautics and Space Administration, 2017).

[8] The deal is valued at $373.5 million if the agency exercises its option for all five Soyuz seats. Boeing obtained the seats from Russian company RSC Energia as part of a settlement of a legal dispute between the two companies involving Sea Launch. Boeing won a judgment of more than

though NASA-Roscosmos cooperation for transporting astronauts to and from the ISS could in principle be over by 2020, the very configuration of the ISS operations makes the conduct of autonomous activities barely possible. In order to ensure safe and continuous operations on-board the ISS, Russian and the U.S. astronauts and scientific communities will hence be obliged to continue interacting at least until 2024.

What is however much more remarkable than this continued cooperation on the ISS was the recent announced signature of a joint statement by Roscosmos and NASA about a possible new partnership for human exploration of the Moon and deep space in the post-ISS context. The statement was signed on 27 September 2017 in the margins of the 68th International Astronautical Congress (IAC) in Adelaide, Australia, although behind-the-scenes discussions between the two agencies had been taking place throughout the year. The agreement covers at the moment only research studies for deep space exploration that "reflects the common vision for human exploration" of the two agencies, but that research, according to the joint statement could eventually "support joint development of the Deep Space Gateway (DSG), a human-tended facility in cislunar space that NASA has proposed as a technology and operational test bed for future human missions to Mars" (Foust, 2017c). In a subsequent statement, Roscosmos announced that the two agencies had already reached an understanding on the "international technical standards of the docking … for the establishment of the station in the near-moon orbit." According to Roscosmos, the partners also discussed the possibility of using a Russian super heavy rocket to complete construction of the orbital moon station, in addition to the SLS (Weitering, 2017).[9]

A few weeks after this landmark announcement, Russian representatives partic-ipated in the so-called Checkpoint Review of the DGS concept, which was held in Houston from 12 to 14 October 2017 to discuss possible contributions of the ISS partners to the DSG. One of the concepts prepared by the Russian delegation was the construction of a Lunar Mission Support Module (LMSM), which "would be attached to the DSG, in the lunar orbit to give the cislunar outpost extra capabilities in life-support, storage and berthing the spacecraft involved in the exploration of the lunar surface". The proposed Russian contributions to the DSG—currently renamed Lunar Orbital Platform-Gateway (LOP-G) in NASA's proposal for the 2019 federal budget—were intended "to address already known limitations in the current design of the DSG to handle the wide variety of missions considered within the project. Notably, the Russian delegation, led by cosmonaut Sergei Krikalev, also included a representative from the international cooperation office of RKK Energia, which had previously criticised potential Russian involvement on the DSG as "an unnecessary distraction on the road to the lunar surface" (Zak, 2017b).

$320 million against Energia from a federal court in May 2016, but subsequent legal filings indicated that the two companies were negotiating a settlement.

[9]"At the first stage, [the deep-space gateway] is supposed to use the American superheavy SLS in parallel with the domestic heavy rockets Proton-M and Angara-5 M," Roscosmos officials said. "After the creation of the Russian superheavy rocket, it will also be used [for] the lunar orbital station" (Ibid).

All in all, while the prospect of NASA and Roscosmos jointly building the LOP-G are still far from becoming concrete, (the LOP-G is indeed still in conceptual phase without commitment to resources or commitment to making it a programme from the political establishment), what this increasing Russian and American convergence underscores is the recognition of the mutual necessity to continue cooperating in the space domain. On the Russian side, Moscow's strategic decision to join the NASA-led LOP-G in spite of the changes this would cause to its original human lunar exploration plans more specifically stems from the realisation that cooperation with the U.S. is a *conditio sine qua non* to fulfilling its future space ambitions, advancing the capabilities of Russian industry to a higher degree of maturity and retaining the status of major spacefaring nations in the international space arena. From the American perspective too (at least at NASA level), there is the recognition that any large exploration programme cannot take place without Russian critical contributions (e.g. the LMSM, which would deliver critical elements for ensuring the success of the LOP-G concept). And this is even more so if taking into account that NASA is barred from cooperating with the other major player in the field of human spaceflight, China, and that future European contributions are not expected to be substantial.

All in all, several elements pinpoint that Russia and the United States seem bound to cooperate also in the future space exploration context. And this is not only because Russian-U.S. interdependence in space is symptomatic of mutual dependence—and hence vulnerability—but also, and perhaps more importantly, of mutual benefit.

3.2.2 Russia and China—Back-up Partners?

One of the most relevant axes of Russia's international space diplomacy has been the relationship with China. Cooperative links were first established in the mid-1950s, when the USSR provided the Chinese with two R-1 missiles and technical assistance for the development of its missile programme. Although suspended in August 1960, as a result of the political tensions that culminated in the Sino-Soviet split, cooperation was restored immediately after the collapse of Soviet Union, and has grown tremendously since then, contributing to the fast-pace development of China's space programme.

Following the signature of the first 10-year intergovernmental agreement on space cooperation in December 1992, the Chinese were invited to study the Soyuz spacecraft, Russian ground and tracking facilities, and environmental control systems for manned spacecraft (Harvey, 2007). In March 1995 a new intergovernmental agreement was signed, specifying Russian assistance to China in human spaceflight and the sale of engines. The Chinese bought Russian RD-120 rocket engines, and later an entire life support system, a Kurs rendezvous system, a docking module, an entire Soyuz capsule—emptied of equipment and electronics—and a Sokol spacesuit.[10]

---

[10]The Russians however refused to sell RD-170 engines, a powerful LOX/Kerosene engine originally used for the first stage of Energya. Ibid.

The agreement also included the training of two Chinese astronauts in Star City, and the opportunity for 20–50 Chinese specialists to attend the training, which took place from 1996 to 1998 (Harvey, 2004). Despite this strong assistance from Russia, relations in the field of human spaceflight have typically remained that of a "buyer-seller", with no active participation of either country in the other's human spaceflight programmes.

In May 2000, cooperation was further institutionalised through the establishment of the Space Cooperation Sub-committee during the Russian and Chinese prime ministers' meeting. The Sub-Committee has held regular meetings ever since, with "two multiannual cooperation agreements… adopted, a first five-year one from 2001 until 2006 and a second ten-year one running from 2007 until 2016" (Mathieu, 2008). The two agreements have identified over 20 cooperation areas, including

(a)   Earth Observation
(b)   space science (e.g. Ultraviolet Space Observatory, joint system of radio inter-ferometers, Spektr UF, Radioastron, etc.)
(c)   deep-space exploration (including Mars exploration with the Russian Phobos Grunt and the Chinese Yinghuo-1).[11]

In contrast, satellite navigation is an area in which China and Russia did not come to terms, at least in the first decade of the 2000s. Whereas at that time China was interested in the development and use of the GLONASS system and Russia in China's financial contribution to revive its navigation system, no agreement was reached because of technology transfer issues and the military stakes of such cooperation. Eventually, Russia opted to reinforce cooperation with India and China to develop an autonomous navigation satellite system (Rathgeber, 2007).[12]

Although in this and other areas, Moscow appeared particularly keen on protecting its technological gap with respect to China—as also demonstrated by the 2007 Moscow court prosecution of Igor Reshetin for transferring classified space-related information to that country[13]—it is also aware that its industry is in danger of losing its strategic edge and that cooperation with China could bring the required resources to make the Russian space industry more innovative, competitive and commercially self-sustainable. As argued by ESPI in the 2012 *Yearbook on Space Policy*, "Russia's industrial difficulties make Russia an ideal partner for China…. China is, of course, a master of industrial production, albeit not yet of the kind of advanced manufacturing required for space, and China would be likely to be keen to get access to Russian

---

[11]This mission, launched in November 2011 from the Baikonur Cosmodrome, resulted in a failure.

[12]In January 2006, Sergei Ivanov, declared India "Russia's only cooperation partner in GLONASS" (Ibid).

[13]In December 2007, the Moscow court sentenced Igor Reshetin, the chief executive of Tsniimash-Export, a producer of rockets and missiles working closely with the Russian Space Agency, to 11.5 years in prison for passing dual-purpose technology to China. "The other three defendants in the criminal case were sentenced to five to 11 years. Investigators said Reshetin and his co-accused had transferred know-how that could be used to design nuclear missiles to the China Precision Machinery Import-Export Corporation, causing losses to Russia of 110 million roubles" (RIANOVOSTI, 2007).

space technology for domestic production, something which seems to have happened quite a bit in China's current human spaceflight programme. Russia certainly has a very realistic view of China, and visa-versa, both are keen to leverage space, and neither is clearly inferior to the other in the overall scheme of things. From a partnership perspective, an auspicious configuration!" (Hulsroj, 2014).

In addition to these programmatic rationales, cooperation could rest on shared political objectives, namely the resolve to team-up against the hegemonic interest of the U.S., be it in space or on Earth. Notably, Moscow and Beijing presented a joint diplomatic initiative to the Conference on Disarmament (CD), with the submission of a "Draft Treaty on the Prevention of the Placement of Weapons in Outer Space" in February 2008.[14] The draft treaty can be read as a way for both the actors to oppose alleged U.S. space dominance, but for Russia it was also an instrument to prevent a dangerous arms race, whose costs could not be sustained.

3.2.2.1 Closer Ties Amid Political Mistrust

With the subsequent deterioration of Russia's political relations with the West in 2014, the U.S. foreign policy posture aimed at containing both Chinese and Russian power projections as well as the situation of NASA (prohibited from cooperating with both China as a result of the Wolf amendment and Russia, following the 2014 sanctions over the latter's annexation of the Crimean Peninsula), cooperation between the two countries has in a sense become a forced choice (Aliberti, 2015).

Remarkably, few weeks after the suspension of cooperation by NASA and the announcement that Russia was not interested in expanding the operations of the ISS beyond 2020, Russia started to trumpet a future of closer ties with the Chinese partner. To illustrate, during a roundtable discussion held at the First Russia-China Expo in Harbin on 30 June 2014, Russian Deputy Prime Minister Dmitry Rogozin stressed: "If we talk about manned space flights and exploration of outer space, as well as joint exploration of the Solar System—primarily it is the Moon and Mars—we are ready to go forth with our Chinese friends, hand in hand" (LaRouche Pac, 2014).

Notwithstanding this clear statement of interest and diplomatic niceties contained in the speech of Dmitry Rogozin, the prospects of closer Russia-China relations were—at least initially—used as a bargaining chip to show the West possible fallouts of their isolation policies, rather than a genuine desire to engage with the Chinese partner. Indeed, traditionally Russia has been visibly reluctant to talk openly about its cooperation with China and has always visualised the bilateral relations under its leadership. It is for instance remarkable that while the Chinese received training at Star City to become *taikonauts*, no Chinese ever flew with the Russians as "passenger", and vice versa (Mathieu, 2008). Imperial mind-sets are after all hard to let go.

---

[14]The draft treaty was submitted together with China and can be read as a way for both the actors to oppose alleged U.S. space dominance, but for Russia it is also an instrument to prevent a dangerous arms race, whose costs could not be sustained by Russia.

More broadly, when looking at the Russian approach to cooperation with its neighbour, a key caveat is their mutual historical mistrust. Indeed, the Chinese-Russian relationship has been in fact "bedevilled by pervasive mistrust, rooted in historical grievances, geopolitical competition and structural factors" (Kotkin, 2009). Chinese officials are well aware of the so-called "China-threat theory" that is deeply rooted among both the general public and Russia's elites and harks all the way back to the Tsarist Empire. In addition, Russia appears extremely wary of Beijing's rising economic and political influence within the international system (Medeiros, 2009). China, in return, has long seen Russia as its voracious neighbour, witness Russia's part in China's century of shame and Mao's statements on Chinese readiness to let hundreds of millions die if necessary to defend the country against Soviet aggression.

Admittedly, at present Sino-Russian relations look to be the most stable bilateral relationship maintained by Russia.[15] The two countries are engaged in an all-dimensional, multi-tiered and wide-ranging cooperation that also extends to military-to-military cooperation and is substantiated in the Shanghai Cooperation Organisation (SCO).[16] While trade relations do not play an important role in the current Sino-Russian interplay—especially when compared to China's profound economic interdependence with the U.S. and Europe—Russia remains an indispensable source of energy supply (in particular oil and natural gas) for Beijing (Aliberti, 2015).

This does not however mean that Russia has more bargaining chips than China. Quite to the contrary, the partnership is inherently "asymmetric", and in China's favour. Stephen Kotkin, professor of history at Princeton University, has underlined that "China extracts considerable practical benefits in oil and weapons from Russia. In return, Beijing has been flattering Moscow with rhetoric about their 'strategic partnership' and coddling it by promoting the illusion of a multipolar world. In many ways, the Chinese-Russian relationship today resembles that which first emerged in the seventeenth century: a rivalry for influence in Central Asia alongside attempts to expand bilateral commercial ties, with China in the catbird seat" (Kotkin, 2009). Also Bobo Lo, a former Australian diplomat in Moscow, has remarkably labelled their strategic partnership an "axis of convenience" primarily pushed by the need to constrain U.S. diplomatic and military power, rather than by any real will to establish a mutually complementary and cooperative relationship (Lo, 2008).

---

[15] As documented by Evan Medeiros, since the late 1980 s, Beijing has made gradual and consistent efforts to upgrade relations, driven largely, but not exclusively, by mutual concerns about U.S. power and the U.S. democracy-promotion agenda. In 1994, China and Russia formed a "cooperative partnership," followed by a "strategic cooperative partnership" in 1996, and the signing of a full treaty on "Good Neighbourliness, Friendship, and Cooperation" in 2001. These agreements led to a series of sustained high-level interactions, which remain the "thickest" part of this bilateral relationship. Since 1996, Chinese and Russian leaders have held annual summit meetings and in 2014 they solidified the relationship with a new strategic partnership agreement that led to important deals in the field of energy and economic cooperation.

[16] Founded in 2001 by China, Russia, Kazakhstan, Kyrgyzstan, Russia, Tajikistan, and Uzbekistan, the SCO is a six-member security association with the objective, at least in Russia's view, of forging a quasi-military alliance that could counter NATO. Within the organisation, cooperation mainly include joint military exercises, intelligence sharing and counterterrorism, but economic and cultural cooperation is also covered.

These considerations do not automatically imply that such an axis cannot become enduring and lead to a "polarisation" of the international community—including the space community—between opposing blocks. In fact, the Sino-Russian relation proves to have evolved and deepened during the past five years, with recent dynamics pinpointing a pathway of true engagement. Because both Beijing and Moscow continue to be isolated—and in a sense contained—by the American grand strategy, the two great powers currently see no other choice than to become closer allies. It is for instance remarkable that in the midst of the standoff over the Ukraine and the launch of U.S. and EU sanctions, in the spring of 2014 the two countries ratified a Strategic Partnership agreement, which has been widely regarded as "the most enhanced in terms of depth and breadth of economic, political, and security relations of any one of China's or Russia's network of strategic partnerships" (Savic, 2016).

Some of the high-profile deals emerging from the 2014 Strategic Partnership have included a 30-year, $400 billion gas supply agreement between Gazprom and China National Petroleum Corporation. The landmark gas supply deal was indirectly referred to in the 2014 Strategic Partnership as a measure aiming to fortify the Sino-Russian energy partnership and hence as a way to demonstrate to the "West" alternative cooperative schemes.[17]

3.2.2.2 Recent Developments

This evolution in the political realm has been suitably reflected also in the space arena. On 30 June 2014—i.e. in the immediate aftermath of the U.S. sanctions that partially suspended NASA-Roscosmos cooperation, Russian and Chinese Deputy Prime Ministers Dmitri Rogozin and Wang Yang announced the signature of a Memorandum of Understanding (MoU) "on establishing a control group for the implementation of eight strategic projects" (The Moscow Times, 2014). Through this MoU, signed in the margins of the Russia-China Space Cooperation Sub-committee during the Prime Ministers' Meeting, the two countries more specifically intended to promote cooperation in deep space exploration, manned spaceflight, earth observation, satellite navigation, space-related electronic parts and components, and other areas.

Among the cooperation fields envisaged in the MoU, particularly indicative of Russia's changed posture in its space diplomacy, was the announced cooperation between GLONASS and China's Beidou system, as it followed the American refusal to host GLONASS ground stations in the U.S. and Russia's decision to disable American GPS stations on its territory. According to the MoU, each country agreed to place three ground stations in the other country, so as to enhance quality in the provision of the respective services (LaRouche Pac, 2014).

Equally indicative of this changed posture was the subsequent announcement about the possible purchase of Russia's RD-180 engines by China, and of Chinese micro radio electronics components by Russia (TASS News Agency, 2016a). Clearly, the two contracts were connected to the ban imposed by the U.S. on the supply to

---

[17]For an analysis on the agreement see: (Koch-Weser & Murray, 2014).

Russia of U.S. spacecraft and devices that used parts made in the U.S., as well as on the ban in the 2015 Defense Authorisation Act on the purchase of Russian engines for the U.S. EELV programme.

During the same period, the two countries also started to engage in more extensive discussions for the renewal of their cooperation agreement (which was set to expire in 2016) with the aim of expanding it into more active cooperation in the field of human spaceflight and robotic space exploration. In this respect, several options had been under considerations and were discussed during the following two years, including:

(a) Cooperation on robotic exploration of the Moon, Venus, Mars, including Russian participation in the Chinese mission to return samples from Mars in the 2020s and Chinese participation in the Venera-D mission
(b) Cooperation on crewed space projects, including the conduct of joint experiments in space medicine and the development of foods for astronauts on-board the upcoming Tiangong space station
(c) Cooperation on launch technology, including supplies of Russia's advanced space rocket engines to China and possible collaboration in the manufacturing of rocket engine technology for the development of super-heavy lift launch vehicles.

Among the three fields of envisaged cooperation, that on launch technology proves to be the most ground-breaking. For one thing, this is an area that could pave the way for ambitious undertakings, including the much-discussed human landing expeditions on the Moon by China. In addition, this is an area in whose development both the Russian and Chinese sides have clear incentives to cooperate. China is currently facing serious issues in the development of its Long March-9 rocket and it is clearly interested in tapping the vast expertise accumulated by Russia in the field of rocket engines. On the Russian side, there is an opportunity to tap into the material resources that China would provide in order to advance its plans for super-heavy launchers, which have been recently facing financial issues and inevitable delays (Zak, 2017d). Indeed, unlike previous cooperative ventures characterised by a "buyer-seller" relationship, the envisaged cooperation could include shared efforts in the development of a new super-heavy lift launcher. As also disclosed to the Russian News Agency TASS by Russian Ambassador to China Andrey Denisov, "the point [of this cooperation] is not to deliver specific equipment but to organize long-term mutually advantageous cooperation of the sides, which are objectively close to each other from the viewpoint of technical and technological compatibility" (TASS News Agency, 2016b).

Remarkably, in the spring of 2016 Russia and China began joint work with "the aim to devise a set of unified standards to be used in manufacturing space technologies, including those necessary for future crewed lunar missions. The countries also aim to develop standards for docking units and electrical connectors" (Nowakowski, 2016). As confirmed by several commentators, such efforts should be viewed in the broader framework of "designing a heavy rocket and establishing interaction in the sphere of space stations and long-distance flights" (TASS News Agency, 2016b). In June 2016, it was also reported that within the mechanism of the Russia-China Space Cooperation Sub-committee during the Prime Ministers' Meeting, the two countries engaged in discussions for reaching an intergovernmental agreement on the protection of Russia's intellectual property rights in the sphere of rocket engines and space technology.[18]

An understanding on these issues was apparently reached at the last meeting of the Russia-China Sub-Committee on Space Cooperation held on 8–10 August 2017. At the conclusion of this meeting the signature of a five-year cooperation agreement for the period 2018–2022 was also announced. As reported by Glavkosmos, the agreement covers five major issue areas, namely: lunar and deep space exploration (both robotic and human), space vehicles and ground infrastructure, hardware components and space materials development, sharing of Earth's remote sensing data, and space debris research (Glavkosmos, 2017a; Xin, 2017)

Irrespective of whether and how these programmes will eventually be implemented, what is important to highlight is the almost simultaneous signature of this cooperation agreement with the joint statement by Roscosmos and NASA on cooperation around the NASA-led LOP-G project. The move is in fact very indicative of Russian space strategy, underscoring Moscow's resolve to maintain the highest degree of freedom in its space diplomacy.

3.2.3 Relations with India—Shrinking Rationales

As in the case of China, Russia's cooperation with India has been key to the advancement of India's space ambitions. Many firsts for the Indian space programme were in fact facilitated by the Soviet Union. The launch of India's very first satellite—the Aryabhata—in the mid 1970s was carried out using a Soviet Kosmos launch vehicle, while almost a decade later a Soviet-led Intercosmos mission put the first, and to this day only, Indian citizen in space. These relations between Moscow and New Delhi, however, evolved in a diametrically opposed direction as compared to Moscow's relations with Beijing. Indeed, whereas in the 1960s Russia and India engaged in a prolific partnership that became even more substantial in the 1970s—i.e. at a time when relations with China were at the lowest level—since the 1990s—i.e. when Sino-Russian cooperation was revived—engagement with India in the space arena has started to witness a progressive decline.

Following the collapse of the Soviet Union, cooperation with India was essentially limited to transfer of cryogenic engine technology for the development of India's heavy-lift launch vehicle, the Geosynchronous Satellite Launch Vehicle (GSLV).[19]

---

[18]Form the Russian perspective, the major issues involved in the implementation of this cooperation were the protection of the rights of intellectual property along with generally-accepted international legal aspects of this activity" (TASS News Agency, 2015a, b).

[19]The plans for the development of this new launcher—named GSLV—were approved in 1987. According to the configuration studies, the GSLV would be a three-stage launcher comprising one solid rocket motor stage derived from the PSLV first stage, one storable liquid stage using the Vikas engine, and one cryogenic stage, in addition to four liquid-fuel engine strap-ons using one Vikas engine each. While much of the technology was designed to employ heavier derivatives of the

But even that limited cooperation eventually faced insurmountable obstacles. Initially, Russia agreed to provide India with the technological know-how India needed to independently launch its satellites into GSO. The Indian Space Research Organisation (ISRO) and Glavkosmos formalised the technology transfer agreement in June 1991, under which the now dissolved USSR would deliver two KVD-1 cryogenic engines and associated technology by 1995 at a very competitive price (just €188 million).[20]

The agreement, however, would never fully see the light, as it immediately prompted the wrath of the U.S., which saw potential for misuse of such technology in connection with India's nuclear weapon arsenal and tensions with neighbouring Pakistan, which also had a nuclear programme. In May 1992, the George Bush administration condemned the agreement as a violation of the Missile Technology Control Regime (MTCR) and imposed sanctions on both ISRO and Glavkosmos that barred U.S. firms from cooperating with them (Aliberti, Aliberti 2018). Initially, the Yeltsin government rejected the U.S. request to cancel the deal, primarily because of the urgent need for hard currency imposed by the collapse of the Soviet Union. However, as detailed by Moltz, "after steady American lobbying and the U.S. delivery of a substantial aid package to Russia at the Vancouver Clinton-Yeltsin summit in April 1993, the Russian government eventually agreed to amend the Indian deal by withholding the cryogenic production technology and selling only the completed boosters themselves" (Moltz, 2012). Accordingly, in July 1993, Russia backed off the initial contract with ISRO, proposing a revised agreement for the transfer of two off-the-shelf engineering models and seven read-to-fly KVD-1 engines. While the new contract was signed, the failed technology transfer contributed to exacerbating India's irritation vis-à-vis both the U.S. and Russia, and its resolve to proceed on its own for the development of cryogenic technology.[21] In retrospect, this U.S. blockade of the initial Indo-Russian technology transfer contract of 1992 effectively cost ISRO 25 years of technological advancement (as ISRO introduced the GSLV-Mk

---

PSLV rocket, the planned introduction of a 12-tonne cryogenic, liquid-hydrogen powered engine for the third stage would prove a major technological feat for India. According to Indian engineers it would have taken up 15 years of work before the country could fully master cryogenic technology. Understandably, India initially sought to acquire such technology from abroad.

[20]The KVD-1 was an old engine developed in 1964 by the Isayev Design Bureau for the USSR lunar landing programme. It was test-fired in 1967, but it was never used as the Moon landing programme was cancelled.

[21]Between 1997 and 2000 the seven KVD-1 engines were delivered to India, enabling the first set of flights with the GSLV, starting in 2001. The first attempt was made on 28 May 2001 (though it failed), while the first operational flight was successfully carried out in September 2004. However, at the same time decisions were taken favouring accelerated indigenisation, particularly in view of the GSLV I's underperformance. Setbacks in the GSLV programme as a consequence of four subsequent launch failures between 2006 and 2010 triggered a fundamental review of the entire GTO launch service programme. While theoretically one final GSLV Mk I launch could have been carried out using the remaining Russian upper stage in stock, decisions had been made to not use it any longer, de facto ending the GSLV Mk I programme.

III launcher equipped with an Indian-built cryogenic engine only in 2017) and also created serious burdens on the previously close ties between Roscosmos and ISRO (Aliberti, 2018).

3.2.3.1 Bold Plans, Small Results

Relations between the two respective space programmes seemed, however, to recover momentum in the course of the 2000s, greatly supported by the solidification of the broader political relations that followed the signature of a strategic partnership agreement by Russian President Putin and Indian Prime Minister Vajpayee in 2000.[22] Ground-breaking cooperation plans were envisaged by the respective space agencies in three key domains, namely:

(a) satellite navigation
(b) robotic space exploration
(c) human spaceflight.

Because these areas were entirely new domains for India, the envisaged cooperation even fed the impression that Russia could become New Delhi's most dependable partner in space. As detailed below, however, all these attempts to cooperate ended rather disappointingly.

(a) In the area of navigation, Russia and India had signed two cooperation agreements on GLONASS between 2005 and 2007, envisaging Indian launches of GLONASS-M satellites on-board India's GSLV, as well as joint development of the future GLONASS-K satellites and of users' equipment (Mathieu, 2008). Through these agreements India sought to gain valuable expertise in the area of satellite navigation as well as preferential access to data, while Russia intended to reduce the financial burdens associated with the redeployment of GLONASS and to make India's augmentation system reliant on GLONASS rather than the American GPS. None of the agreements, however, came into fruition, with Russia eventually deciding to launch the GLONASS satellites on its own rockets from Plesetsk. The motivation for terminating the cooperation was never clarified, but it can be argued that, as in the case of China, Moscow appeared keen on protecting its technological gap with respect to India and thus to prevent the possible transfer of sensitive technologies. Nevertheless, in 2011 Russia signed an agreement with India to provide India's military with preferential access to its PNT data (Korovkin, 2017).

(b) The envisaged partnership on space science and exploration turned out to be equally ill fated. Besides the utilisation of Russian ground stations to support

---

[22]The main areas of cooperation covered by this partnership are energy and defence (Mathieu, 2008). Closer alliances with influential Asian partners such as China, Turkey and India were vital to the health of the Russian economy at the time, and more recently somewhat inevitable given the sanctions of Western economic powers on Russia in connection to the annexation of Crimea in 2014.

India's first exploration mission to the Moon, cooperation in this field foresaw the launch an Indian payload on-board the Russian Coronas-Photon mission in 2009, and Russian participation in ISRO's second lunar mission Chandrayaan-2. In the first case, IRSO provided an RT-2 gamma-telescope to investigate the processes of free energy accumulation in the sun's atmosphere, but the mission failed less than one year after the launch. As for the Chrandrayaan-2 lunar rover mission, Russian participation consisted in providing the landing module building on the state-of-the art technologies developed for the Phobos-Grunt mission. However, as detailed by Vladimir Korovkin, "the Russian party did not provide the landing module in time, rescheduling the delivery first for 2013 and then for 2016. [Although some] cited the failure of Fobos-Grunt mission in 2011 to be part of the reason, though the mission failed due to an unsuccessful launch and never managed to test the landing device" (Korovkin, 2017). Ultimately, in 2015 India decided to reschedule the mission in 2018 but to follow through without Russian participation, and so continued on its increasingly autonomous path with regards to its space programme.

(c) A third area of cooperation that Russia and India envisaged during the 2000s was human spaceflight. India had decided to embark upon an autonomous manned spaceflight programme following the landmark achievements reached by China in 2003. The programme was given the green light by ISRO in 2006 and initial funding began in April 2007, with the objective of sending the first Indian astronaut into orbit by 2014 (Peter, 2008). To support this ambitious programme, ISRO took the decision to human-rate its upcoming Geo-Synchronous Launch Vehicle (GSLV-Mk III) and develop a two-person manned capsule. Actions were initiated accordingly. By the middle of 2007, ISRO had validated its re-entry technology with the successful recovery of a space capsule, and had started to work on pre-projects, including long-lead items for human missions such as spacesuits and simulation facilities (Peter, 2009). Given the invaluable expertise of Russia in this domain, ISRO also sought support from Roscosmos. Following a working group on space cooperation with Russia, an agreement was reached between the two countries in December 2008 for the Indian manned spacecraft to be built following the trusted Russian Soyuz design, thus echoing the same path implemented with China.[23] In that context India also considered sending one of its citizens into space on board a Russian spacecraft to acquire the skills necessary for future manned space missions, and expressed interest in participating in the development of a new Russian manned spacecraft.

Also in this case, however, the plans never materialised. In part because of this failed cooperation and in part because of the much-delayed development of the GSLV-Mk III (a delay that can be once again attributed to the missed transfer of cryogenic engine technology by Russia in the 1990s), the human spaceflight programme in India had to be postponed indefinitely.

---

[23] Also echoing the Chinese approach, it was proposed that the first mission would be for a day, while the second for a week (Harvey, 2004).

3.2.3.2 Weaker Rationales and Selective Cooperation

From an overall perspective, it is safe to argue that all these disappointing cooperation experiences have certainly contributed to a diminished interest in closer cooperation, at least from the Indian perspective. Not surprisingly, little interaction has taken place over the past few years between Roscosmos and ISRO. Remarkably, not only have the two partners grown apart, but they have also started to compete in the commercial arena, particularly in the launch service market. India today is considered a potentially prolific commercial launch provider, as proven in their 2017 single launch of 104 mostly foreign satellites, a situation that now contributes to threaten Russia's established position in this market (Aliberti, 2018).

It is also important to highlight that, with the exception of few areas, at the moment both partners have no particular needs nor real opportunities to work together on space projects, because of limited rationales at programmatic level. India has become a space power of its own and has in many ways grown self-sufficient, thereby reducing the need for extensive cooperation with the Russian neighbour. In the area of navigation, for instance, India has developed its own navigation satellite system—the Indian Regional Navigation Satellite System (IRNSS), also referred to as NAVIC. Similarly, in the area of robotic space exploration India can now boast increasing expertise, while in the field of access to space it is putting the finishing touches on its own cryogenic engine technology for its new launch vehicle, the GSLV-Mk III. While cooperation in human spaceflight could still come under consideration, also in this arena India has decided to develop crucial technology in a gradual—yet autonomous—manner.[24]

A possible exception is given by the telecommunications sector. For instance, it was reported that in August 2017 a delegation from Glavkosmos visited ISRO's Satellite Centre in Bangalore to discuss "the Russian bid on a fully electrical propulsion system equipped with stationary plasma thrusters for the future geostationary communication satellites of ISRO. The bid had been drafted specially for the ISRO Satellite Centre and issued for the initial bidding phase" (Glavkosmos, 2017b). Irrespective of the eventual success of this specific initiative, it must be noted that this is also an area where Russian players have to compete with the commercial efforts undertaken by other spacefaring nations as well as with the indigenous solutions coming from ISRO.

---

[24]ISRO's focus is currently on four major lines of activity: (a) development of new technologies required for human spaceflight in the areas of crew module systems including re-entry and recovery elements, environmental control and life support systems and flight suits, and crew escape system; (b) unmanned flight-testing of the crew module systems; (c) validation of the performance of the crew escape system through pad abort tests; (d) development of environmental control and life support Systems (ECLSS) and carrying out an integrated test of ECLSS with the crew module on the ground. As contended by Unnikrishnan Nair, Project Director of the Human Spaceflight initiative, "the successful completion of this phase of critical technology developments will enable ISRO to prove its capability leading to the full-fledged HSP" (Nair, 2015).

If the autonomy gained by India is now diminishing the rationale for cooperation from a programmatic perspective, the same applies to India's increasingly close relations with the U.S. Indeed, in the eyes of Moscow one of the main drivers for cooperating with India was to promote a "more democratic" international environment. In other words, space cooperation was intended as a political tool to promote deeper political engagement and counterbalance the prevailing hegemony of the U.S. in the international system by promoting a multipolar international system. Hence, the recent placing of India among the closest allies of the U.S. and the quasi-alliance set under Obama administration has removed much of the political rationale that had been underpinning Russia-India space cooperation.

All in all, while the defence and energy ties between Russia and India remain strong (Busvine & Pinchuk, 2016),[25] especially so in light of Western sanctions against Russia, space cooperation appears to be on decline in comparison to the close ties currently pursed by Russia with China.

This, however, cannot be taken to imply that there are no longer avenues for cooperation between Moscow and New Delhi. Indeed, some space-related cooperation remains for instance viable in connection with the defence industry, an area where Russia can still leverage its valuable expertise to engage in mutually beneficial undertakings. In this respect, it is worth underlining that in an effort to boost India's defence exports, BrahMos Aerospace has been chosen to be India's commercial arm for Indian defence products sold abroad. BrahMos is partly owned by the Indian Defence Research Development Organization (DRDO) and partly by the Russian NPO Mashinostroyeniya, the former of which retains a majority share with 50.5% equity. The idea of having a separate entity—namely BrahMos—handling international commercial activity for the DRDO came as a result of increased global demand for Indian products after a series of exhibitions and air shows, and aims to skyrocket Indian defence exports to US$2 billion in two years. However, gaining significant market share over the medium to long-term may prove difficult considering the dominance of well-established industry giants (Sputnik News, 2017a). Just days after news of the newfound partnership broke, India successfully test launched the BrahMos Extended Range supersonic missile off its eastern coast. Considered a milestone step for Indian defence, it now extends India's striking capability well beyond its current 180 miles (around 290 km). Complete with a guidance and stealth technology, the two-stage missile reached 2.8 Mach during the test flight. Touting BrahMos' as the global pinnacle of supersonic cruise missile system providers, great optimism and confidence seems to be spreading throughout the Indian defence industry in light of recent achievements (Maass, 2017).

Just a month prior to the announcement of the BrahMos partnership, in February 2017 India and Russia agreed to fortify Indian defence capabilities through a GLONASS ground station in India. The increased navigation accuracy and coverage means better wartime command for Indian forces and secondarily brings with it civil applications. The vast Russian defence technology used by Indian armed forces is

---

[25] In 2016 several energy and defence deals worth billions of dollars were closed at the 2016 annual Indo-Russian summit (Ibid).

compatible with GLONASS, thus making this a sensible move from a technological perspective. It is not too far-fetched to consider that this improved navigation capability by the addition of GLONASS signal to the already operational IRNSS was a key move to bolster technological capabilities in preparation for the partnership with supersonic cruise missile company BrahMos (Aliberti, 2018).

3.2.4 Relations with Other Actors and Groupings

In addition to its evolving ties with the major spacefaring nations, over the past ten years Russia's space diplomacy has sought a clearer definition of its relations with key countries within the Commonwealth of Independent States (CIS) and to extend its reach with emerging spacefaring nations in the Asian context, in the Middle East and Latin American regions, as respectively exemplified by recent cooperation with South Korea, Iran and Brazil.

3.2.4.1 The CIS: Ukraine

Ukraine has occupied a special place in Russia's external relations. After the collapse of the Soviet Union, Ukraine inherited between 15 and 30% of the Soviet space potential and support infrastructure. It inherited a particularly large share of the testing capabilities of the former Soviet Union as well as relevant technological and industrial assets (Mathieu, 2007). A key example is given by Yuzhnoye, a production and development facility for ballistic missiles and launch vehicles inherited by Ukraine.

The loss of Ukrainian space capability was for a long time considered of relatively minor consequences for Russia, given the overall positive relations between the two countries, and the fact that Ukraine continued to seek close cooperation with Russia for the pursuit of its space activities. By the late 1990s, two joint ventures had been created among Russian and Ukrainian companies, mostly to allow for continued development and operations of the Dnepr and Zenit launch vehicles.

- the exploitation of Dnepr had been entrusted to the International Space Company (ISC) Kosmotras, a joint venture created in 1997 with the participation of Russian, Ukrainian space agencies and industrial companies. Unlike other joint ventures, the launcher system was of mixed origin, with Ukrainian partners being responsible for design supervision over the Russian converted ICBM and providing the payload adaptor.[26]

---

[26]ISC Cosmotras was entrusted with the development and commercial operation of Dnepr-1. Roscosmos thus partnered with the Ukrainian space agency and a group of industrial companies from both Russia and Ukraine in this company.

- the exploitation of Zenit, which comprised roughly 70% Russian and 30% Ukrainian components,[27] was managed first by Sea Launch through a subcontract granted to Space International Services (SIS)—which was initially in charge of mission integration and launch operations (Sea Launch maintained management and quality control functions)—and then directly by SIS, as Sea Launch withdrew from this business in 2013 (see Annex D).

In addition to these joint ventures, Ukraine has continued to supply Russia with important hardware components for its space programme and the two countries have cooperated in some satellite programmes and scientific research.

However, as the recent political developments in the Crimea started to affect the broader Russia-Ukraine relationship, cooperation began to face insurmountable hurdles. Since 2014, Russia has initiated a costly effort to replace Ukrainian components in its space programme. Some of the imported hardware has turned out to be not as simple as envisaged. For example, more than a billion roubles (€14.4 million as of March 2016) had to be allocated for organizing production of four types of titanium tanks at the Voronezh Mechanical Plant.[28]

In addition, Russia openly turned away from the use of Ukrainian vehicles by excluding their use for Federal Space Programme launches and putting into question the continued cooperation on ICBM reconditioning. Inevitably, the joint ventures suffered severely from this situation, as the vehicles themselves required components from both countries. In the case of Zenit, no acceptable solution for nationalisation by one or the other country could be found and the programme was consequently suspended in 2014. The acquisition of Sea Launch by the S7 group now allows to restart Zenit exploitation even though it appears likely that instead of using a Zenit vehicle, Sea Launch would opt for using a new exclusively Russian vehicle. In the case of Dnepr, a buyout of Ukrainian interests and subsequent transfer of Ukrainian responsibilities to a Russian company was to facilitate renewed exploitation, even though a resumption of activities appears unlikely today. More broadly, as Ukrainian leadership increasingly leans westwards in its foreign policy, the overall prospects of Russia-Ukrainian cooperation remain rather bleak.

3.2.4.2 The CIS: Kazakhstan

Another crucial—yet sometimes burdensome—CIS partner of Russia in its space endeavour has been Kazakhstan. This was obviously so because, following the collapse of the Soviet Union, Russia lost access to its traditional launch site, Baikonur, now situated within the borders of a newly independent state. The Kremlin hence had to negotiate an agreement with the government of Astana, under which Russia

---

[27]SIS partners included Ukrainian Yuzhnoye and Yuzhmash, and Russian Energia, KBTM and TsENKI.

[28]Tanks designed to contain highly pressurized helium gas are used by pneumatic systems of rocket engines on such rockets as Proton, Angara and Briz upper stage. As it turned out, they had been manufactured exclusively by a production plant at KB Yuzhnoe in Dnepropetrovsk, Ukraine.

was granted a long-term lease for a land surface of 7650 km^2 at an annual fee of US\$115 million until 2050.

Even though Kazakhstan also needed (and still needs) Russian presence to pursue its space activities due to its lack of domestic launch capability, some disputes between the two countries have periodically arisen, with Astana's government regularly putting into question the lease agreement for Baikonur or using its prerogative to halt the granting of launch authorisations.

To illustrate, the terms of use of Baikonur require the submission of an annual launch schedule the preceding year, which has led to launch delays for missions, not included in such a list. On occasion, the total number of launches approved has also been lower than that requested by Russia. In addition, Kazakhstan has repeatedly criticised environmental damages caused by failed launches, leading to compensation claims in 2007 and 2013 as well as temporary withdrawal of launch authorization for Proton and Dnepr in 2007. Furthermore, there have been serious disputes also on debris drop zones. Sometimes, these have led to the interruption of Russian operations from Baikonur or led to delays in the launches of Dnepr following recurrent denials of launch authorisation by both Kazakhstan and Uzbekistan.[29]

It is however important to stress that as Russia began demonstrating its determination to carry out launch operations away from Kazakhstan onto Russian territory, Kazakhstan began to ease its posture. In fact, in 2013 the re-established Russian-Kazakh intergovernmental commission on the Baikonur Cosmodrome decided that Kazakhstan would no longer prevent Russia from launching Proton vehicles in return for reducing parts of the land leased to Russia. The two countries also agreed to gradually discontinue leasing space facilities exclusively to Russia and consider a joint use and operations of the cosmodrome. The latest agreement, signed at the end of 2016 between the Ministry of Defence and Aerospace Industry of Kazakhstan and Roscosmos SC, includes an eight-year road map until 2025 and establishes a protocol defining steps to be taken in case of incidents. Against the background of warming Russia-Kazakhstan relations, Baikonur was announced to become a site for commercial launch services, while Vostochny is to be used for the federal programmes.

As of 2017, the space relations between the countries seemed to further improve with the revival of the old collaboration on the Baiterek project—a joint Russian-Kazakh initiative for building an ecologically friendly launch complex to support the operations of the Russian Angara launch vehicle. Cooperation on this project started in December 2004, when Moscow and Astana signed a first agreement establishing the Russia–Kazakhstan Baiterek Joint Venture, in which each country held a 50% stake. In 2010, however, the project was put on the backburner due to insufficient funding, issues related to Kazakh's entry in the Baikonur cosmodrome and, most importantly, Russia's intention to use Angara from its Vostochny cosmodrome. With the new Russian plans to develop the new Phoenix (Soyuz-5) medium-range rocket, the plans were eventually revived within the above-mentioned 2016 agreement. According to the new roadmap, funding for the Baiterek project shall start in 2019,

---

[29]In 2014, a 10-months Kazakh ban on Russian oil imports in 2014 led to shortage of heating fuel at Baikonur and lack of hot water availability at the launch site.

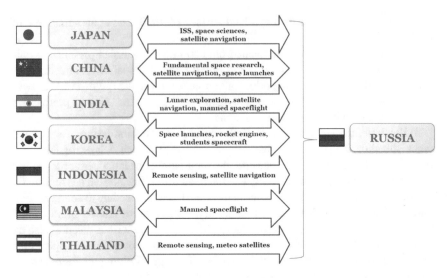

Fig. 3.2 Russia's regional space cooperation in Asia (*credit* Volynskaya, 2014)

while the development and manufacture of equipment, construction and installation should begin in 2021–2023. The integrated flight tests are slated for 2024–2025, with the first launch of the Phoenix launcher from the Baiterek Rocket and Space Complex scheduled for 2025 (Dyussembekova, 2017).

3.2.4.3   The Asian Context

Outside the CIS framework, over the past 10 years Russia has sought to extend its reach within the Asian context. Besides the close cooperation with China, Roscosmos has established cooperative links with Japan, South Korea, Indonesia, Malaysia and Thailand. An overview of these cooperation projects is provided in Fig. 3.2 (Volynskaya, 2014).

Within the Asian regional context, particularly remarkable are Russia's ties with Japan and the Republic of Korea, especially if one considers the very close relationships that both Tokyo and Seoul have with Washington. It is also of interest that whereas Russian relations with the Japan Aerospace Exploration Agency (JAXA) have been mainly framed in the multilateral ISS programme, with the Korean Aerospace Research Institute (KARI) cooperation has involved even sensitive technology transfers and joint development of the Korean Space Launch Vehicle-1 (KSLV-1). As explicitly acknowledged by KARI, Russia's selection as cooperation partner in the field stemmed from the fact that it was "one of the leaders in the development of space launch vehicles. In fact, Russia was the only advanced country in the field of space launch vehicles that was willing to cooperate, and had the intention to commercialize Korea's launch vehicle technology" (Korea Aerospace Research Institute, 2016).

The bilateral agreement for the joint development of the launch vehicle (including the assembly and testing of the system) and its related infrastructure was signed by KARI and Khrunichev in 2004. According to the agreement, the first stage of the Naro development project was to be undertaken by Russia, while the second stage was to be carried out by KARI (Ibid). The agreement included a provision allowing for the use of Russian Angara technology for a maximum total of three KSLV-1 launches. The first two launches of KSLV-I, carried out in 2009 and 2010 respectively, resulted in failure. The agreement foresaw a third launch in case of failure of any of the first two launch attempts with delivery of a stage at Russian expenses in case the failure could be linked to the stage. While the causes of the failure remained disputed, a third launch was carried out on 30 January 2013, and it was successful (Kyle, 2013). Interestingly, the launch also allowed Russia to use this KSLV-I flight to qualify its own Angara stage, clearly underpinning Russia's ability in maximising its cooperation with foreign countries.

More recently, Russia's Centre for the Operation of Space Ground-Based Infrastructure (TsENKI) has been selected to assist KARI in the further expansion of its national Naro Space Centre, including an upgrading of the launch pad to support multipurpose satellites and the launch of the KSLV in the future. In exchange, "Russia's ground-based signal correction station for the GLONASS satellite navigation system could be placed in South Korea" (Sputnik News, 2017b).

For the time being, Korea will possibly continue to rely on foreign launchers, including Russian Soyuz for the launch of its payloads into orbit, as evidenced by the signature, in August 2017, of two launch contracts with Glavkosmos Launch Services for the delivery of Korea's CAS 500-1 and CAS 500-2 Earth observation satellites (Glavkosmos, 2017c).[30] Notwithstanding these currently close relations, space relations between Russia and South Korea continue to be offset by a high degree of disquiet, as evidenced in the past by issues such as Roscosmos' decision to replace South Korean first astronaut Ko San just one month before his spaceflight on Soyuz, as well as the numerous difficulties encountered during the development of Naro-1, among others (Krasnyak, 2017).

3.2.4.4   The Middle East Context: Iran

In the Middle East region, the most important space partner the Russian government has tried to nurture close relations with is Iran. Iran is a strategic player in Russia's geo-politics, and science diplomacy, including space cooperation, has served to both reinforce political ties with Teheran and achieve programmatic objectives of the Russian space programme, namely a market for Russian products.

---

[30]The primary purpose of the CAS500-1 and -2 satellites is to provide panchromatic and multi-spectral image data with the Advanced Earth Imaging Sensor System-Compact (AEISS-C) payload. CAS500-1 was developed by KARI; CAS500-2 by the Korea Aerospace Industries (KAI) (Ibid).

Space cooperation began to unfold in the 2000s, when Tehran turned scientific development into a national priority by, inter alia, establishing "a number government bodies in charge of priority research, i.e. the Iranian Space Agency (ISA) the Special Office for Nanotechnologies, and the National Science Foundation for supporting outstanding researchers and specialists" (Demidenko, 2014). Russo-Iranian cooperation has since then concentrated in the areas of remote sensing and navigation defence missiles. In 2005, for instance, Russia built and launched Iran's first satellite, the Sina-1 Earth Observation satellite, and in 2007, it signed a contract for the delivery of S-300 air-defence missiles to Tehran.

In June 2010, however, the then Russian president Dmitry Medvedev temporarily suspended all cooperation commitments in light of the UN Security Council's Resolution 1929 against the development of the Iranian nuclear programme (RBC, 2010). Space-related cooperation was then renewed in 2014 after the sanctions were lifted. On 10 April 2014, Iran signed a protocol of cooperation with Roscosmos on a wide range of activities including training Iranian astronauts in Russia, the possible manufacturing and launch of Earth observation and telecommunication satellites for Iran, as well as the granting of access to the data of Russia's Resurs-DK and Resurs-P satellites (RT News, 2014). In turn, Iran agreed to place the elements of the Differential Correction and Monitoring System (SDCM), as well as a quantum optical system, on its territory to maintain the Russian GLONASS navigation system (Sputnik News, 2014).

Cooperation further expanded in 2015, when on the occasion of the 12th Russian International Aviation and Space Show (MAKS), the Russian-Iranian High-Level Commission co-chaired by Dmitry Rogozin (the Deputy Prime Minister of Russia) and Sorena Sattari (vice president for science and technology) was set up (TASS News Agency, 2015a, b). At the same event, Bonyan Danesh Shargh, an Iranian aerospace company, signed a preliminary agreement with Russian companies NPK BARL and VNIIEM for building a remote-sensing system based on an upgraded version of the Kanopus-V1 observation satellite. According to the agreement, NPK BARL will build and adapt the system's ground infrastructure, and VNIIEM will be responsible for building and launching the satellite with a Russian Soyuz carrier rocket and putting it into the Earth's orbit by 2018. According to *SpaceWatch Middle East* the remote sensing satellite that Iran and Russia agreed to build will likely be the National Remote Sensing Satellite (NRSS) that the Iran Space Agency announced, along with the National Communications Satellite (NCS) (SpaceWatch Middle East, 2016).

In October 2016, Moscow announced it had completed the delivery of the S-300 systems to Iran, despite the concern expressed by the U.S. and Israel (RT News, 2017). For the moment, Iran remains in search of more serious investment and technology improvement. So far Russia is not ready to provide the full help but considers Iran to be a promising partner in future and is engaging in the competition for the Iranian market. In September 2017 president Putin announced the new goal for the GLONASS system—the increase of data quality up to 3.5 m in 2020 that requires building additional ground infrastructure and can become one of the drivers for future cooperation between Russia and Iran (VEDOMOSTI, 2017). On a note of caution, it should however be highlighted that possible space cooperation between Russia and Iran's arch-rival Saudi Arabia that is currently under discussion may put some strains on future Russo-Iranian space ties (Barrabi, 2016).

3.2.4.5 The Latin American Context: Brazil

In the South American context, Russia's outreach has been more recent as compared to other regions and has been driven by a mix of programmatic, economic and political needs. To illustrate, Russia has reached an agreement with both Nicaragua and Brazil to host a monitoring station for its GLONASS satellite navigation system in their countries, as a way to both ensure larger penetration of its PNT services and strengthen its political influence in the region. Remarkably, out of the eight GLONASS stations located outside Russia, four are in Brazil (The Moscow Times, 2015).

Apart from the GLONASS support, Russia has been interested in conducting joint space launches from Brazil's Alcantara Launch Centre as well as in the development of small and medium-lift carrier rockets (Sputnik News, 2017c). In this respect, an agreement on the evolution of the VLS-1 rocket featuring the incorporation of a Russian liquid upper stage was concluded in May 2011 between Roscosmos and the Brazilian Space Agency.

Against the background of Russia's deteriorating relations with Ukraine, it was reported that in 2015 the Brazilian Presidency was preparing to terminate the strategic partnership with Ukraine for the development of Cyclone-4 (which was suffering from numerous delays), also in the hope of improving political ties with Russia within the BRICS framework. In April 2015, the Brazilian Space Agency announced the formal termination of the Cyclone-4 programme by the government, also in light of the necessary—yet still missing—export control agreement with the U.S. By the same time, both the U.S. and Russia were competing for a strategic role in supporting Brazil in its plan to launch commercial satellites from an equatorial base. However, the NSA espionage scandal of 2014 and U.S. concerns about Brazil's partners getting access to the top U.S. technologies eventually led to significant complications in the development of space cooperation between Washington and Brasilia.

The situation thus seemed to be developing in favour of Russia, especially following the last meeting between Vladimir Putin and Brazilian President Michel Temer in June 2017. On that occasion, President Putin specifically stressed that the two countries were committed to closely cooperate in the field of access to space, remote sensing and GLONASS ground stations. Temer, in his turn, expressed interest in expanding the network of such stations and in expanding Brazilian space capabilities through the support of Russia.

3.2.4.6   Multilateral Cooperation: The BRICS Framework

The changing geopolitical situation of recent years has pushed Russia's space diplomacy. to explore new multilateral cooperation formats in order to advance both political and programmatic objectives. The most noticeable development in this respect has been the space cooperation established within the Brazil-Russia-India-China-South Africa (BRICS) association.

From a political perspective, Moscow sees "its participation in the BRICS group as one of the fundamental directions of its long-term global strategy" (Okouneva, 2012). BRICS cooperation has in part replaced the concept of a strategic triangle between the three Eurasian giants that was put forward by former Russian Prime Minister Y.M. Primakov in 1998 and re-launched by Russia's Ministry of Foreign Affairs in 2007, but with limited results.[31] As the Kremlin visualises it, in light of their combined political, economic and demographic weight, the BRICS countries have enough power to promote a "more democratic" and multipolar international order where the hegemonic interests of the U.S. are constrained. In addition. given the lukewarm relations between China and India, Russia can play a pivotal role in this grouping. Equally important, the core "principles shared by the BRICS (and stated in March 2012 at the five countries' summit in New Delhi), such as the necessity for overhauling the international and economic financial system, these countries' right and proper role in the world economy as well as the principles of non-alienation, from the free choice of priorities in international policy, the refusal of force, to diktat and coercion—all that corresponds to the general direction of Russian foreign policy and fulfils the international positioning of the country's interests" (Okouneva, 2012).[32]

In this context, space cooperation has become an important tool to promote closer ties. Possible multilateral space cooperation has been discussed over the past few years at the various BRICS summits. At their annual meeting in July 2017, the space agencies of the five BRICS countries eventually agreed to create a shared Remote Sensing Constellation. It is currently envisaged that the BRICS Remote Sensing Satellite Constellation will be implemented in two phases: "phase 1 would create a remote sensing data sharing system, making the data from each of the member countries' existing EO satellites available to all the other members as well. Phase 2, which will be further discussed and defined in the near future, will involve the creation of a new EO satellite constellation" (Campbell, 2017). The transition from a virtual constellation to the effective creation of a multinational constellation featuring joint planning of new satellite missions will be agreed upon in future BRICS summits.

---

[31]This is mainly because neither China nor India has seemed particularly interested in the idea, given the potential negative rifts such an alliance could generate in their respective foreign policies, whose liberty of manoeuvre and cooperative interplay with Western countries would become compromised.

[32]Russia has "economic and commercial ties with all the BRICS, it takes part in all activities including summit meetings, those of the Foreign, Finance, Economy, and Agriculture Ministers, not to mention meetings of High Representatives for security matters, heads of BRICS states' personal representatives and their subordinates and the G-20 etc. The BRICS' full agenda is fully accepted and approved in Russia" (Ibid).

In addition to remote sensing, the five BRICS space agencies are currently planning to further space cooperation in the field of navigation, space science and possibly even human spaceflight (Drozhashchikh, 2017). At the moment, however, this novel cooperation format remains at an early stage.

3.3 Assessing the Trajectory of Russia's International Posture

The recent evolution of the Russian space sector and the recent strains in Russia's economic performance and international relations (e.g. Ukraine, Crimea, Syria) have led to an evolution of its cooperation with foreign partners. While Russia has generally remained very open to international cooperation in its space programmes, its attitude towards its traditional partners in the space field has changed. More specifically, the evolving political context has triggered a sensible reduction in cooperation activities with several former partners and a striving towards greater autonomy from foreign sources. This tendency intensified following the 2014 Crimea crisis, which marked the end of the historical cooperative undertakings that Russia put in place in the launcher sector in the mid-1990s with CIS and Western countries. In addition to disengaging from multinational joint-ventures, and repatriating their shares to Russian entities, Russia put a stop to future use of vehicles not entirely manufactured domestically and has started to accelerate the construction of a new launch site on its territory to replace its dependence on Baikonur. More broadly, it has started to develop production capabilities for critical components that are not yet domestically available.

In parallel with this aspiration towards technological non-dependence, Moscow has been trying to reinforce its relations with other partners, particularly emerging spacefaring nations such as Brazil, South Korea and Indonesia, and non-Western countries such as China and Iran, in the form of both technology transfer and joint activities. As emerges from the analysis, such diversification in the partnership portfolio has been in part driven by Russia's resolve to counterbalance the increasing strains on cooperation with the West, (and indeed to mitigate the risks of isolation in Russian international outreach in space), and in part, it has been driven by the increasing needs to find alternative sources of revenues caused by the budget reductions of recent years. Such diversification of Russia's partnerships may bring important developments in the international space arena, but they should not be overstated either.

Russia's past and current cooperation experience with countries such as India, China or Brazil shows that relations with these countries face inherent hurdles, including a mismatch in interests, priorities and capabilities, as well as general political mistrust. As the failed cooperation initiatives with India in the field of launchers, space exploration and human spaceflight clearly show, not all desirable partnerships might actually materialise, and in some instances Russia's position in the international space arena may even border isolation. The fact "that so little is happening in alliance building has both Russian and non-Russian reasons, but Russia as a teacher in India, South Korea, Brazil, China will not be sustainable without rapid renewal of its industry, and it is virtually impossible to see how this will come about" (Hulsroj, 2014). Russian credibility is on the descent, and this is well understood by Russia's partners and possible partners.

In addition, some of the most recent evolutions in Russia's external relations (such as for instance the envisaged participation in the LOP-G framework and the recently established cooperation with ESA for lunar exploration) show that Western countries remain to date obliged partners for Russia in its space endeavour.

References

Aliberti, M. (2015). *When China goes to the moon....* Vienna: Springer.

Aliberti, M. (2018). *India in space: Between utility and geopolitics*. Vienna: Springer.

Aliberti, M., & Tugnoli, M. (2016). *European launchers between commerce and geopolitics*. Vienna: European Space Policy Institute.

Barrabi, T. (2016, June 16). *Russia builds space program with Saudi Arabia, Iran Amid Roscosmos' Struggles*. Retrieved January 9, 2018 from International Business Times: http://www.ibtimes. com/russia-builds-space-program-saudi-arabia-iran-amid-roscosmos-struggles-1969236.

Busvine, D., & Pinchuk, D. (2016, October 15). *India, Russia agree to missile sales, joint venture for helicopters*. Retrieved May 20, 2017 from Reuters: https://www.reuters.com/article/ us-india-russia-helicopters/india-russia-agree-to-missile-sales-joint-venture-for-helicopters-idUSKBN12F058.

Campbell, K. (2017, July 4). *Brics bloc agree remote sensing space constellation project*. Retrieved January 20, 2018 from Engineering News: http://www.engineeringnews.co.za/article/brics-bloc-agree-remote-sensing-space-constellation-project-2017-07-04.

Demidenko, S. (2014, October 28). *Russia-Iran Scientific and technical cooperation: Is there any?* Retrieved December 9, 2017 from Russian International Affairs Council: http://russiancouncil. ru/en/analytics-and-comments/analytics/russia-iran-scientific-and-technical-cooperation-is -there-an/.

Drozhashchikh, E. (2017, September 7). *Space platform for BRICS cooperation*. Retrieved January 20, 2018 from Russian International Affairs Council: http://russiancouncil.ru/en/analytics-and-comments/analytics/space-platform-for-brics-cooperation/.

Dyussembekova, Z. (2017, March 16). *Baiterek rocket and space complex set to launch in 2025*. Retrieved December 1, 2017 from The Astana Times: https://astanatimes.com/2017/03/baiterek-rocket-and-space-complex-set-to-launch-in-2025/.

Foust, J. (2017a, January 17). *NASA considering Boeing offer for additional Soyuz seats*. Retrieved October 2, 2017, from Space News: http://spacenews.com/nasa-considering-boeing-offer-for-additional-soyuz-seats/.

Foust, J. (2017b, February 28). *NASA signs agreement with Boeing for Soyuz seats*. Retrieved October 2, 2017 from Space News: http://spacenews.com/nasa-signs-agreement-with-boeing-for-soyuz-seats/.

Foust, J. (2017c, September 28). *NASA and Roscosmos to study Deep Space Gateway*. Retrieved October 2, 2017 from Space News: http://spacenews.com/nasa-and-roscosmos-to-study-deep-space-gateway/.

Gibbons, G. (2014, June 17). *GNSS monitoring stations slide into U.S.-Russia rift*. Retrieved October 8, 2017 from Inside GNSS: http://www.insidegnss.com/node/4067.

Glavkosmos (2017a, August 11) *Glavkosmos continues to develop the Russia-China space cooperation.* Retrieved November 2, 2017 from Glavkosmos: http://glavkosmos.com/news/glavkosmos-continues-to-develop-the-russia-china-space-cooperation/.

Glavkosmos. (2017b, August 30). *Glavkosmos discussed a promising project with Indian partners.* Retrieved October 1, 2017 from Glavkosmos: http://glavkosmos.com/news/glavkosmos-discussed-a-promising-project-with-indian-partners/.

Glavkosmos. (2017c, August 21). *Signature of launch services contracts with KARI & KAI.* Retrieved January 15, 2018 from Glavkosmos: http://glavkosmos.com/news/signature-of-launch-services-contracts-with-kari-kai/.

Harvey, B. (2004). *China's space program. From conception to manned spaceflight.* New York: Springer.

Harvey, B. (2007). *The rebirth of the Russian space program. 50 Years after Sputnik, New Frontiers.* Chichester: Springer. (Praxis).

Hulsroj, P. (2014). The psychology and reality of the financial crisis in terms of space cooperation. In C. Al-Ekabi et al. (Eds.), *Yearbook on space policy 2011/2012. Space in times of financial crisis.* Vienna: Springer.

Koch-Weser, I., & Murray, C. (2014). *The China-Russia gas deal: Background and implications for the broader relationship.* Washington D.C.: U.S.-China Economic and Security Review Commission.

Korea Aerospace Research Institute. (2016). *First Korea space launch Vehicle Naroho (KSLV-I).* Retrieved January 12, 2018 from KARI: https://www.kari.re.kr/eng/sub03_03_02.do.

Korovkin, V. (2017). Evolution of India-Russia Partnership. In R. P. Rajagopalan & N. Prasad (Eds.), *Space India 2.0; Commerce, policy, security and governance perspectives* (pp. 246–262). New Delhi: Observer Research Foundation.

Kotkin, S. (2009). The unbalanced triangle. *Foreign Affairs, 88*(5), 130–138.

Kramer, M. (2014, April 4). *NASA policy to suspend contact with Russia 'unprecedented,' But maybe symbolic, expert says.* Retrieved October 31, 2017 from SPACE.com: https://www.space.com/25360-nasa-russia-policy-unprecedented-symbolic.html.

Krasnyak, O. (2017, July 18). *South Korean-Russian space cooperation: Mistrust & national diplomatic styles.* Retrieved January 12, 2018 from USC Center on Public Diplomacy: https://uscpublicdiplomacy.org/blog/south-korean-russian-space-cooperation-mistrust-national-diplomatic-styles.

Kuchins, A., & Zevelev, I. (2012). Russia's contested national identity and foreign policy. In H. R. Nau & D. M. Ollapally (Eds.), *Worldviews of aspiring powers: Domestic foreign policy Debate in China, India, Iran, Japan and Russia.* New York: Oxford University Press.

Kyle, E. (2013, March 7). *KSLV.* Retrieved November 13, 2017 from Space Launch Report: http://www.spacelaunchreport.com/kslv.html.

LaRouche Pac. (2014, July 1). *China and Russia continue to Deepen space cooperation.* Retrieved October 29, 2017 from LaRouche Pac: http://archive.larouchepac.com/node/31183.

Lo, B. (2008). *Axis of convenience: Moscow, Beijing and the new geopolitics.* Washington D.C.: Brookings Institution Press.

Luzin, P. (2014, May 16). Space station wars: The empire strikes back. *Russia Direct.*

Maass, R. (2017, March 14). *India test fires BrahMos extended range missile.* Retrieved from Space Daily: http://www.spacedaily.com/reports/India_test_fires_BrahMos_Extended_Range_missile_999.html.

Mathieu, C. (2007). *Space in Central and Eastern Europe. Opportunities and challenges for the European space endeavour.* Vienna: European Space Policy Institute.

Mathieu, C. (2008). *Assessing Russia's space cooperation with China and India: Opportunities and challenges for Europe.* Vienna: European Space Policy Institute.

Medeiros, E. S. (2009). *China's international behaviour. Activism, opportunism and diversification.* Santa Monica: RAND Corporation.

Miller, J. (2014, June 2). *Russia to 'restrict' US-run GPS satellites.* Retrieved October 8, 2017 from BBC News: http://www.bbc.com/news/technology-27662580.

Molotov, I., Voropaev, V., & Borovin, G. (2016). *Recent developments within the ISON project.* Vienna: UNOOOSA.

Moltz, J. C. (2012). *Asia's space race. National motivations, regional rivalries, and international risks.* New York: Columbia University Press.

Musa, A. (2016, November 1). *Russia and America: Crisis on the planet, friendship in space.* Retrieved November 2, 2017 from Russia Direct: http://www.russia-direct.org/analysis/russia-and-america-crisis-planet-friendship-space.

Nair, S. U. (2015). Initiatives on India's human spaceflight. In P. M. Rao (Ed.), *From fishing hamlet to red planet.* New Delhi: Harper Collins.

NASA Office of Inspector General. (2017). *Nasa's commercial crew program: Update On development and certification efforts.* NASA Office of Inspector General: Washington D.C.

National Aeronautics and Space Administration. (2014, April 2). *NASA internal memo: Suspension of NASA contact with Russian entities.* Retrieved October 9, 2017 from SpaceRef: http://spaceref. com/news/viewsr.html?pid=45536.

National Aeronautics and Space Administration. (2017, March 10). *NASA studying shared venus science objectives with Russian Research Institute.* Retrieved November 13, 2017 from NASA: https://moon.nasa.gov/news/2017/03/10/nasa-studying-shared-venus-science-objectives-with-russian-space-research-institute/.

Nowakowski, T. (2016, July 23). *Russia and China envision joint space exploration.* Retrieved October 31, 2017 from Spaceflight Insider: http://www.spaceflightinsider.com/organizations/ roscosmos/russia-china-envision-joint-space-exploration/.

Okouneva, L. (2012). *Russia and Brazil in the BRICS group—Future ambitions.* Brussels: European Laboratory of Political Anticipation.

Oznobishchev, S. (2014, May). New cold war, new space race. *Russia Direct, 10.*

Peter, N. (2008). Developments in space policies programmes and technologies throughout the world and Europe. In K.-U. Schrogl, C. Mathieu, & N. Peter (Eds.), *ESPI yearbook 2006/2007: A new impetus for Europe* (p. 96). *Springer, Vienna.*

Peter, N. (2009). Developments in space policies programmes and technologies throughout the world and Europe. In K.-U. Schrogl, C. Mathieu, & N. Peter (Eds.), *ESPI yearbook on space policy 2007/2008: From policies to programmes* (p. 81). Vienna: Springer.

Rathgeber, W. (2007). *China's posture in space. Implications for Europe.* Vienna: European Space Policy Institute.

RBC. (2010, June 22). *Д.Медведев запретил поставки в Иран ракетных комплексов С-300 (D. Medvedev banned the supply of missile systems S-300 to Iran).* Retrieved January 5, 2018 from RBC: https://www.rbc.ru/politics/22/09/2010/5703de6e9a79470ab502527e.

RIANOVOSTI. (2007, December 3). *Reshetin sentenced for 11.5 years for passing technology to China.* Retrieved October 29, 2017 from RIANOVOSTI: http://en.ria.ru/russia/20071203/ 90747889.html.

RT News. (2014, May 7). *Russia to train Iranian cosmonauts, build recon sats—report.* Retrieved January 7, 2018 from RT News: https://www.rt.com/news/157496-russia-iran-space-satellite/.

RT News. (2017, March 4). *Iran successfully tests Russia-supplied S-300 anti-aircraft system— media.* Retrieved January 15, 2018 from RT News: https://www.rt.com/news/379427-iran-tests-russian-s300/.

Savic, B. (2016, December 7). *Behind China and Russia's 'special relationship'.* Retrieved November 2, 2017 from The Diplomat: https://thediplomat.com/2016/12/behind-china-and-russias-special-relationship/.

Senske, D., & Zasova, L. (2017, January 31). *Venera-D: Expanding our horizon of terrestrial planet climate and geology through the comprehensive exploration of venus.* Retrieved December 13, 2017 from National Aeronautics and Space Administration: https://solarsystem.nasa.gov/docs/ Venera-D_Final_Report_170213.pdf.

SpaceWatch Middle East. (2016, December 19). *Iran urges Russia to start building remote sensing satellite.* Retrieved January 18, 2018 from SpaceWatch Middle East: https://spacewatchme.com/ 2016/12/iran-pushes-russia-start-building-remote-sensing-satellite/.

Sputnik News. (2014, May 13). *Iran to host Russian satellite navigation facility*. Retrieved January 12, 2018 from Sputnik News: https://sputniknews.com/world/20140513189786057-Iran-to-Host-Russian-Satellite-Navigation-Facility/.

Sputnik News. (2017a, March 9) *BrahMos aerospace to be Indian DRDO's commercial wing abroad*. Retrieved from Space Daily: http://www.spacedaily.com/reports/BrahMos_Aerospace_to_Be_Indian_DRDOs_Commercial_Wing_Abroad_999.html.

Sputnik News. (2017b, September 26). *S. Korea to develop Naro Space Center in cooperation with Russia—Roscosmos*. Retrieved January 10, 2018 from Sputnik News: https://sputniknews.com/world/201709261057695351-south-korea-russia-space-iac/.

Sputnik News. (2017c, June 21). *Putin: Russia, Brazil consider joint space launches from Brazilian spaceport*. Retrieved January 9, 2018 from Sputnik News: https://sputniknews.com/science/201706211054855598-putin-russia-brazil-space-launches/.

TASS News Agency. (2015a, July 6). *China may buy rocket engines for its space program from Russia—Deputy Prime Minister more:* http://tass.com/economy/806267. Retrieved October 31, 2017, from TASS News Agency: http://tass.com/economy/806267.

TASS News Agency. (2015b, November 18). *Rogozin: Iran and Russia will develop cooperation in the field of space*. Retrieved January 12, 2018 from TASS News Agency: http://tass.ru/kosmos/2450581.

TASS News Agency. (2016a, June 20). *Russia, China to sign deal paving way to rocket engine contract—Deputy PM more:* http://tass.com/economy/883484. Retrieved October 30, 2017 from TASS News Agency: http://tass.com/economy/883484.

TASS News Agency. (2016b, June 21). *Russia, China may cooperate in developing heavy space rocket—Diplomat*. Retrieved October 30, 2017 from TASS News Agency: http://tass.com/science/883685.

The Moscow Times. (2014, May 19). *Russia, China sign space exploration agreement*. Retrieved October 30, 2017 from The Moscow Times: https://themoscowtimes.com/articles/russia-china-sign-space-exploration-agreement-35569.

The Moscow Times. (2015, September 16). *Brazil and Russia boost space cooperation with new Glonass station*. Retrieved January 21, 2018 from The Moscow Times: https://themoscowtimes.com/articles/brazil-and-russia-boost-space-cooperation-with-new-glonass-station-49594.

VEDOMOSTI. (2017, September 19). *Путин потребовал довести точность «Глонасс» до GPS (Putin demanded to bring the accuracy of "Glonass" to GPS)*. Retrieved from VEDOMOSTI: https://www.vedomosti.ru/technology/news/2017/09/19/734484-putin-tochnost-glonass-gps.

Volynskaya, O. (2014). Recent developments of space policy in the Russian Federation. *International Workshop "Space policy and law for social development in Asia"*. Tokyo: Roscosmos.

Weitering, H. (2017, September 27). *NASA and Russia partner up for crewed Deep-Space Missions*. Retrieved October 2, 2017 from SPACE.com: https://www.space.com/38287-nasa-russia-deep-space-gateway-partnership.html.

Xin, L. (2017, August 31). *China, Russia to have smooth space cooperation, says expert*. Retrieved November 2, 2017 from Space Daily: http://www.spacedaily.com/reports/China_Russia_to_Have_Smooth_Space_Cooperation_Says_Expert_999.html.

Zak, A. (2017b, October 31). *Russia proposes Lunar Mission support module for deep space gateway*. Retrieved November 2, 2017 from Russianspaceweb: http://www.russianspaceweb.com/imp-lmsm.html.

Zak, A. (2017d, December 11). *Russian plans for a super-heavy rocket*. Retrieved December 12, 2017 from Russianspaceweb: http://www.russianspaceweb.com/superheavy.html.

Chapter 4
Prospects for Europe

In this final part, the book examines the possible consequences of these domestic and international developments for Europe. More specifically, this chapter will first assess the current Europe-Russia interplay in the political sphere and then recent and current relations in the space arena. Subsequently, it will evaluate the potential impact of recent changes on Europe's cooperation with the Russian partner. Finally, different options for cooperation between Europe and Russia will be defined and assessed in order to maximize opportunities, while minimising risks.

4.1 Europe-Russia Relations: The Background

Russia is Europe's largest neighbour and due to this geographic proximity, historical legacy, and cultural links, Russia holds a specific position for Europe and vice versa. In the past decades, the relationship between the two sides has evolved from one of reciprocal hostility during the Cold War to a partnership in the 2000s, and again into a stance of political confrontation since 2013. Even though each side has been working towards strengthening political and economic ties, their relations have never fully met each partner's expectations and the recent evolution and posture of both actors has posed serious strains on a relationship that, nevertheless, remains today highly interdependent and well-structured in a variety of domains and markets.

4.1.1 Political Relations: From the PCA to the Five-Principle Russia Policy

In addition to well-developed bilateral relations with some of the European Union (EU) Member States (most notably, Germany) Russia has developed strong relations with the EU over the past 25 years. The basis for EU-Russia relations is the Partnership and Cooperation Agreement (PCA) that was signed in 1994 and came into force in December 1997 for an initial duration of 10 years. Since 2007 it has been

© Springer International Publishing AG, part of Springer Nature 2019
M. Aliberti and K. Lisitsyna, *Russia's Posture in Space*, Studies in Space Policy 18,
https://doi.org/10.1007/978-3-319-90554-9_4

Table 4.1 EU-Russia common spaces

Common space	Objectives
Common economic space	Conditions for increased and diversified trade and creation of new investment opportunities by pursuing economic integration and regulatory convergence, market opening and infrastructure development
Common space of freedom, security and justice	Facilitate ease of movement between EU and Russia within a context free of terrorist threat, organised crime and corruption
Common space of external security	Cooperate on security and crisis management in order to address global and regional challenges, including terrorism and climate change
Common space of research and education	Create and reinforce bonds between the EU and Russian research and education communities

renewed on an annual basis. The PCA established the framework for regular consultations between both parties, including bi-annual summits of heads of states and governments, including the President of the European Commission (EC), the Head of the EU Council and the Russian President. The PCA has been complemented by sectoral agreements covering a wide range of policy areas, including political dialogue, trade, science and technology, education, energy and environment, transport, and prevention of illegal activities (European External Action Service, 2017).

At the Saint Petersburg Summit of May 2003, both partners decided to reinforce their partnership in the framework of the PCA by creating four "common spaces" (European Commission, 2004), as detailed in Table 4.1.

In May 2005, the Moscow Summit adopted objectives and roadmaps for the implementation of those common spaces (European External Action Service, 2005). Remarkably, space cooperation was mentioned several times in these roadmaps as part of both the dialogue on industrial policy and the creation of a common EU-Russia Information Society area within the common economic space.[1]

---

[1] Under the heading "Space", the roadmap more specifically states that the objective is "to build an effective system of cooperation and partnership between the EU and the Russian Federation in the following fields of space activities: Access to Space: Launchers and Future Space Transportation systems; Space Applications: Global Navigation Satellite Systems (GNSS); global monitoring by satellites and satellite communications; Space exploration and the use of the International Space Station (ISS); Space Technologies Development". In terms of actions to be undertaken the document lists: (a) Political cooperation to create favourable framework conditions in the field of space transportation, accompanying the cooperation between the European Space Agency and Russia; cooperation for the development of infrastructure for the launch of Russian SOYUZ-ST Launcher from the European Spaceport in the Guiana Space Centre; (b) Enhance and strengthen cooperation on Galileo and GLONASS GNSS including on compatibility and interoperability between the two systems and the creation of the conditions for industrial and technical cooperation, in the context of an intergovernmental agreement; (c) Provide appropriate environment for fruitful cooperation on Global Monitoring for Environment and Security (GMES) programme and for joint projects

By 2008, the two sides had also begun negotiations for the renewal of the PCA and the possible introduction of a free trade area and free visa travel. However, the Russian-Georgian war in the summer of that year caused a cooling of relations and temporary deferral of the policy dialogue. While the cooperation talks were subsequently resumed, the relationship never fully recovered, but on the contrary started to progressively deteriorate. To a large extent, this stemmed from the adoption by the Russian leadership of a more confident posture on the international scene, which openly criticised the West and Europe, for instance, on NATO's expansion or on the establishment of U.S. missile defence infrastructure in Europe. This assertiveness in advancing national interests has alarmed some European capitals (particularly in Eastern Europe) and led to an unfavourable political context to develop the partnership.

EU-Russia relations entered a new and particularly difficult phase after 2013, when the EU negotiated association agreements (AAs) with Ukraine, Moldova, Georgia and Armenia, envisaging closer economic integration through the creation of free trade areas (FTAs). The Kremlin, which saw the move as an intrusion in its regional sphere of influence, pressured "Ukraine and Armenia to pull out of the association agreements and join its own economic integration project, the Eurasian Economic Union (EEU) instead" (Russel, 2016).[2]

After the decision to go ahead with the signing of its AA with the EU by Ukraine's pro-European leadership, Europe-Russia relations abruptly worsened, with Russia reportedly fomenting separatist uprisings in the eastern Ukrainian regions of Donetsk and Luhansk and eventually annexing Crimea in March 2014. In response to such aggression in Crimea and Eastern Ukraine, in July 2014, the EU adopted a series of sanctions on Russia:

- diplomatic sanctions, including the suspension of EU-Russia summits with indefinite effect;
- economic sanctions targeting Russia's financial, defence and energy sectors;
- individual sanctions imposing travel bans/asset freezes on select entities and individuals;
- sanctions against Crimea, which included a ban on trade and investment between the EU and the contended peninsula.

Russia retaliated with counter-sanctions banning around half of EU agricultural food imports (e.g. fruit, vegetables, meat and dairy) and also waging an aggressive

---

in satellite communication systems; (d) Coordinate the EU and the Russian positions towards the Global Earth Observation initiative (GEO); (e) Continue cooperation and partnership in joint initiatives in space explorations, including Space Science on which relevant activities shall be carried out in the framework of the Common Space on Research and Education, including Culture; (f) Continue cooperation in the use of the ISS; (g) Support joint programmes and projects in Space Technology Development; (h) Setting up an expert group to establish an EU-Russia Dialogue on Space; (i) In the framework of the PCA institutions, establish a mechanism for cooperation to comply with the objectives agreed by the Parties; (j) Exchange information and ensure consultation on respective space programmes" (European External Action Service, 2005).

[2]Whereas in January 2015 Armenia decided to join Russia, Belarus and Kazakhstan in the EEU, Ukraine preferred to proceed with an AA with the EU.

Fig. 4.1 President Putin and EU high representative Federica Mogherini. *Credit* RaiNews 24

1. Full implementation of the Minsk agreements

- Lifting economic sanctions against Russia only after the full implementation of the Minsk I and II agreements is ensured

2. Closer ties with Russia's neighbours

- Pursuing closer relations with the former Soviet republics in the eastern neighbourhood of the EU (including Ukraine) and in central Asia

3. Strengthen EU resilience to Russian threats

- Becoming more resilient to Russian threats such as energy security, hybrid threats, and disinformation

4. Selective engagement with Russia

- Engaging selectively with Russia on a range of foreign policy issues, including cooperation on the Middle East, counter-terrorism, climate change. Some scientific research (e.g. Horizon 2020) is also excluded.

5. Support for people-to-people contacts

- Increasing support for Russian civil society and promoting people-to-people contacts, given that sanctions target the regime rather than the Russian people

Fig. 4.2 The EU's five-principle policy towards Russia

information war against EU countries. With the inevitable deadlock over Ukraine and the subsequent military intervention of Russia in Syria, tensions have remained very high in the past three years (Fig. 4.1).

In response to this standstill and potentially counterproductive uncoordinated actions among Member States, in March 2016, EU foreign ministers agreed with the EU High Representative for Foreign Affairs and Security Policy, Federica Mogherini, on five guiding principles for EU-Russia relations. These principles are detailed in Fig. 4.2.

Despite some differences in views, the Five-Principle Policy outlined by High Representative Federica Mogherini was discussed and endorsed by EU Heads of State and Government at the European Council meeting in October 2016.

The various resolutions approved by the European Parliament between 2015 and 2016 have been consistent with the European Council's posture in all five areas. However, as also highlighted by a report of the European Parliament, "implementing each of these principles faces major difficulties. The EU is unlikely to lift sanctions against Russia while implementation of the Minsk agreements remains stalled; the EU's Eastern Neighbourhood remains a zone of confrontation; EU security is threatened by dependence on Russian energy imports and the destabilising effects of aggressive propaganda; cooperation on international issues has become a victim of tensions between the two sides; repressive legislation obstructs EU support for Russian civil society; and EU-Russian people-to-people contacts are in decline" (Russel, 2016).

4.1.2   Economic Relations: Important Partners

Sanctions have generated painful economic costs for both the EU and Russia. According to the estimates of the European Commission and the Kremlin, the annual cost of sanctions and counter-sanctions is at 0.25% of GDP for the EU (European Commission, 2015), and 2% of GDP for Russia (VEDOMOSTI, 2016). There are also high political costs, as sanctions are one of the main obstacles to normalisation of EU-Russia relations.

The fact that sanctions have come with high costs for both Russia and the EU shows how interdependent the two actors have become. Indeed, despite political tensions, economic ties between the two sides remained close. The EU is by far Russia's largest trading partner and the most important investor in Russia, whereas Russia is the EU's fourth trading partner, but also the EU's leading energy supplier.[3]

As reported by the EC, trade between the two economies showed sharp growth rates until 2012, when it reached record levels. In that year—also thanks to the EU's support—Russia joined the WTO, further expanding opportunities for economic relations with the EU and other foreign partners. However, problems remained with the Russian implementation of the WTO commitments, which has had an impact on further growth (European Commission, 2017). Indeed, bilateral trade declined already in 2013 and it went further down in 2014 against the backdrop of economic difficulties in Russia, Western sanctions, and trade restrictions introduced by Russia on food imports. The strong depreciation of the Russian ruble, the fall in global commodity prices, and the protracted economic recession in Russia led to a further contraction in bilateral economic relations.

---

[3]When looking at Europe's dependency on Russian gas, or Russia's dependency on European money, it has been argued Russian over dependence on energy exports (energy accounts for nearly two-thirds of Russia's total exports), particularly those to the largest energy market in the world (the EU), make Russia more dependent on the current energy relationship than the other way around.

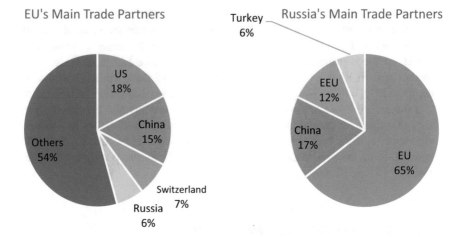

Fig. 4.3 Main trade partners 2015. *Source* European Commission (2017b)

Table 4.2 Trade in goods 2014–2016, €billions

Year	EU imports	EU exports	Balance
2014	182.4	103.2	−79.2
2015	136.4	73.7	−62.7
2016	118.7	72.4	−46.2

Source European Commission (2017)

Table 4.3 Trade in services 2013–2015, €billions

Year	EU imports	EU exports	Balance
2013	14.1	31.1	17.0
2014	12.4	29.8	17.4
2015	11.7	25.3	13.6

Source European Commission(2017)

Russia, which accounted for 10% of the EU's trade in 2012, accounted for less than 6% (and 4% of EU exports) four years later in 2016. Russia and the EU still remain important trade partners, as shown in Fig. 4.3.

In 2016, EU exports to Russia totalled €72.4 billion, while EU imports from Russia amounted to €118.7 billion. The EU trade deficit with Russia was therefore €46.2 billion in 2016, which is primarily the result of significant EU imports of energy products from Russia. Tables 4.2 and 4.3 show the most recent evolution of EU-Russia trade in goods and services.

EU's exports to Russia are dominated by machinery and transport equipment, electrical and electronic equipment, and plastics. EU's imports from Russia are dominated by raw materials, in particular, oil (crude and refined) and gas (see Fig. 4.4).

The EU, in addition, has remained the largest investor in Russia. It is estimated that up to 75% of Foreign Direct Investment stocks in Russia come from EU Member

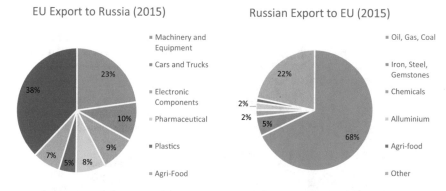

Fig. 4.4 EU-Russia trade composition. *Source* European Commission (2017b)

Table 4.4 Foreign direct investment, €billions

Year	Inward stocks	Outward stocks	Balance
2014	56.4	162.7	101.2
2015	60.9	171.8	111.0

Source European Commission (2017b)

States. The total stock of foreign direct investment in Russia originating from the EU and vice versa is shown in Table 4.4.

4.1.3 Issues and Prospects

Even though economic sanctions have now restricted access to Western capital markets, thus making it difficult for Russian businesses to finance investments, this "economic inter-dependence of supply, demand, investment and knowledge resulted in numerous joint commitments to maintain good economic relations", despite the tense political interplay (European External Action Service, 2017). In addition, given the lack of progress in the implementation of the Minsk agreements and the heavy cost of sanctions, it comes as no surprise that there has been growing objections to their continuation.

Some national governments (such as Italy, France, Austria, and Germany) have expressed cautious reservations, while others (Greece, Slovenia, Slovakia, Check Republic, and Cyprus) have expressed open opposition to continuing sanctions.[4] Nevertheless, even the countries that have publicly criticised sanctions have thus far emphasised that EU unity comes first, with most EU governments continuing to

---

[4]A key determinant in the development of relations between the EU and Russia will be the position of eastern EU Member States, especially Poland and the Baltic States, which have been quite critical towards Russia and closer EU-Russia cooperation.

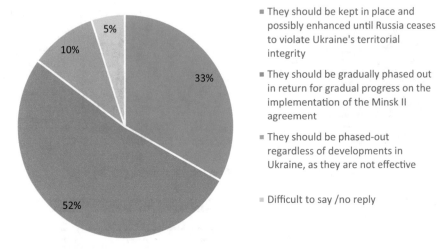

With regard to the EU's sanctions against Russia...

5%
10%
33%
52%

- They should be kept in place and possibly enhanced until Russia ceases to violate Ukraine's territorial integrity

- They should be gradually phased out in return for gradual progress on the implementation of the Minsk II agreement

- They should be phased-out regardless of developments in Ukraine, as they are not effective

- Difficult to say /no reply

Fig. 4.5 Survey on Europe's sanctions. *Source* European Leadership Network (2016)

insist on full implementation of Minsk as a precondition for the normalisation of the economic relations.

All in all, while intra-European unity has hitherto prevailed, it is also clear that it will not be easy to overcome mutual suspicion and the frost in diplomatic relations with Russia. These issues are also reflected in the survey that the European Leadership Network (ELN) carried out with former, present and emerging political, military, and diplomatic leaders from the broader European area with respect to the current state of and future prospects for EU–Russia relations and the pan-European security architecture (European Leadership Network, 2016).[5]

With regard to the sanctions policy, "an overwhelming majority of respondents (85%) remained in favour of maintaining the original linkage of EU sanctions with Russia's actions in Ukraine. Within this group, the majority (52%) supported the position that the sanctions should be gradually phased out in return for progress on the implementation of the Minsk 2 agreements, while a smaller group (33%) remained in favour of maintaining or even enhancing the EU sanctions as a pressure instrument to restore the full territorial integrity of Ukraine, including Crimea" (see Fig. 4.5).

Beyond immediate issues such as the scope and utility of sanctions, "the basic strategic question for the EU seems to be whether to oppose and try to influence Russia's policy by making closer cooperation contingent upon Moscow's adherence

---

[5]The ELN is a non-partisan organisation based in London working to develop collaborative European capacity to address foreign, defence and security policy challenges through its active network of former and emerging European political, military, and diplomatic leaders and through institutional partnerships across Europe. The survey involved 42 respondents represent a diverse and experienced group of individuals from 20 countries and all major regions of Europe.

As a general approach to relations with Russia, the EU should...

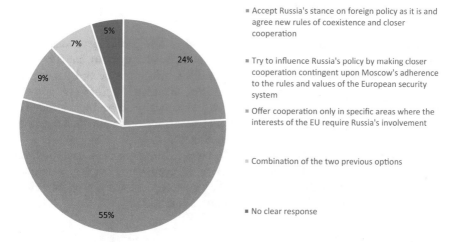

Fig. 4.6 Survey on possible EU's posture vis-a-vis Russia *Source* European Leadership Network (2016)

to the rules and values of the European system, or whether to accept the new Russia and its policy 'as it is' and move towards agreeing the new rules of the relationship" (European Leadership Network, 2016).

According to the survey, a change of course in Russia's policy, rather than accommodation with Moscow, should be the EU's main objective. "The current EU policy of making a full resumption of ties and the deepening of cooperation with Russia contingent on a change in Russian behaviour enjoyed the support of 55% of respondents. However, a mixed group of Western European and Russian participants (24%) expressed the view that the EU should accept that Russia is pursuing a particular kind of foreign policy which is unlikely to change in near future, and should therefore base its policy on a pragmatic rather than transformative agenda" (see Fig. 4.6).

The results of the survey confirm "that there are no simple solutions to improve the state of EU—Russian relations. Apart from an unlikely substantial re-evaluation of Russian policy towards Europe, there seem to be no silver bullets to remove the tensions between the two sides, especially over the common neighbourhood area. Instead of forging a working partnership, the best-case scenario most of the participants can envisage seems to be a gradual and most likely torturous process of establishing a new regime of co-existence and limited EU-Russia cooperation" (European Leadership Network, 2016).

This cooperation will be more likely in areas where the interests of Europe require Russian involvement such as counter-terrorism, Syria and the Middle East, nuclear disarmament and proliferation, energy, global challenges such as climate change and joint scientific research. Remarkably, there are also signs that space can be acknowledged as one of these areas of selective engagement.

4.2   Europe-Russia Space Cooperation: Business as Usual or Selective Engagement?

The recent strains on political dynamics have not directly impacted space cooperation endeavours between Russian and European partners. To a large extent, this is because most space cooperation has been centred on ESA, which has been, by nature of the organisation, detached from political matters. Indeed, as "a European research and development organisation, ESA is a programmatically driven organisation i.e. the international cooperation is driven by programmatic needs more than a general "foreign policy" as is the case for Sovereign States".[6]

However, as emerges from the analysis below, there have been some indirect impacts that have been dictating a selective engagement and a less ambitious approach, which is partially in contrast to the close cooperation and even joint strategy for space activities advocated in the course of the 2000s.

4.2.1   Europe-Russia Space Cooperation Frameworks

Europe and Russia have a long history of cooperation in space, be it at national or pan-European level. European countries and the Soviet Union have cooperated in space for almost 50 years now.

Besides participation in the Intercosmos and Intersputnik programmes by the German Democratic Republic and some of the eastern EU member States, from the 1960s, scientific cooperation was undertaken between the USSR and France, with a first agreement in 1966. At the European level, cooperation between Russia and the European Space Agency (ESA) started in the 1990s and since then it has developed into a strategic partnership. Cooperation that started in space science, was then extended to manned spaceflight and more recently to launchers.

Space cooperation between Europe and Russia has taken place at various levels, in different frameworks and—thematically—in a plethora of fields, including satellite telecommunications, launchers and launch services, manned spaceflight, and science and exploration (Mathieu, 2008).

At national levels, Russia has intergovernmental and interagency agreements, as well as joint working groups and specific activities, with many EU member states. At the European level, all the three-major pan-European institutions involved in space matters, namely ESA, the European Organisation for the Exploitation of Meteorological Satellites (EUMETSAT), and the EU, have entertained cooperative relations with Russia.

- **Cooperation between ESA and Russia**: This started in 1991 with a first Framework Agreement on Cooperation, which enabled joint activities, in particular, in

---

[6]Nevertheless, ESA has supported the EC in the Space Dialogues the EC has set up with United States, Russia, China and South Africa (Fonseca, 2013).

space science and manned spaceflight on the ISS. In the 1990s, ESA established a permanent mission in Russia in 1995 to facilitate dialogue between the two partners and provide information and analysis support, project support and logistic support. The first cooperation agreement was followed in February 2003 by the signing of an "Intergovernmental Agreement on Cooperation and Partnership in the Exploration and Use of Outer Space for Peaceful Purposes", which provides a solid legal basis for cooperation (Mathieu, 2008). Since then, Roscosmos has become—together with the U.S.—ESA's closest international partner.

- **Cooperation between the European Commission (EC) and Russia**: This started in the late 1990s with the involvement of the European Union in space affairs.

  - In December 2001, the EC, ESA and Rosaviacosmos (now Roscosmos SC) signed a joint Memorandum of Understanding (MoU) "New Opportunities for a Euro-Russian Space Partnership" proposing the establishment of a long-term partnership between Russia and Europe in the fields of launchers and the two EU flagship programmes, Copernicus and Galileo. The MoU was followed by an EU-Russia joint statement on space cooperation in May 2002, and by a joint workshop in January 2003 to promote it (European Commission, 2003).
  - Within the framework of the Common Economic Spaces (CES), a tripartite dialogue on space cooperation between ESA, the EC and Roscosmos was established in March 2006 to strengthen cooperation in the fields of Earth observation, satellite navigation, satellite telecommunications, access to space, space science, and space technology developments. Seven joint working groups were created to work on common initiatives, which were to take place mainly through ESA programmes, the EU Framework Programmes and Russia's space programme (Mathieu, 2008).
  - However, policy dialogues with the EC were interrupted following the suspension of EU-Russia summits, and cooperation has been handled directly through ESA.

- **Cooperation between EUMETSAT and Russia**: The Darmstadt-based organisation has mainly cooperated with ROSHYDROMET, which is the institution, within the Russian Federation, in charge of operating satellites with meteorological, ocean and climate-related missions. In a similar way to EUMETSAT, ROSHYDROMET has established a long-term satellite programme that includes geostationary (Electro), low Earth orbit (Meteor) and highly elliptical (Arctica) orbiting satellites. As reported by EUMETSAT, the two organisations started their formal relationship in 1997. The current cooperation framework covers satellite data and product exchange from the respective geostationary and low Earth orbit satellites, satellite validation and calibration, scientific activities and training (EUMETSAT, 2017). Since 2010, a Roshydromet HRPT receiving station in Moscow has been included in the EUMETSAT Advanced Retransmission Service (EARS) network, focusing on regional near real-time data reception. This has resulted in data increase in the coverage of the Northern hemisphere. The current plan is to include a further two stations (located in Novosibirsk and Khabarovsk) in the EARS service in the medium term (Ibid).

Those intergovernmental and interagency agreements at the national and European levels give a sound framework to the cooperation between European companies and research institutes and Russian counterparts.

4.2.2 Sectoral Cooperation Overview

Over the years, Russia and Europe have cooperated in most areas of space activities, and particularly in four main fields, namely human spaceflight, launchers and launch services, space science and exploration, and operational satellite systems (Facon & Sourbès-Verger, 2007).

4.2.2.1 Human Spaceflight

Manned spaceflight figures among the strongest assets of Russia's space activities. Cooperation between Europe and Russia in manned spaceflight has been crucial to enable early participation of Europe in the field and to build and enhance European capabilities in the domain. Cooperation started in the early 1990s with the EuroMir programme, which aimed to offer ESA astronauts experience in space and help all parties to gain experience in international cooperation. The programme saw two missions of European astronauts on-board the Russian Mir, respectively in 1994 and 1996.

Since the late 1990s, Russia and Europe have cooperated within the framework of the ISS, and more specifically with respect to the following elements:

- European astronauts' missions for the launch to, return from, and use of the Russian ISS segment. Those missions have included a complete programme of scientific and technological experiments conducted in the ISS Russian segment. As of 2017, ESA continues to rely only on Russian Soyuz for crew access to the ISS.
- Astronaut training, which continues to be partially carried out in the facilities of Star City, close to Moscow.
- Technical cooperation. The ESA-developed Automated Transfer Vehicle (ATV) used technologies from Russia and has always docked to the Russian port of ISS. Russia contributed to the development of the ATV, in particular, to the design of the docking system. Europe has provided the central data management system of the ISS Russian Service Module and has built the European Robotic Arm (ERA) to be used in the assembly and servicing of the ISS Russian segment. Initially slated for launch in 2015, ERA shall be launched by a Proton rocket in 2018 on the Russian MLM (see further).
- Joint scientific research and experiments conducted in microgravity in particular in the field of biology and medicine.

It terms of science research, ESA has the Institute for Bio-Medical Problems (IMBP) as one of its most reliable partners. Main areas of cooperation are:

- Adaptation of the human body to the conditions of short and long-term space flights, including interplanetary flights;
- Means and techniques of directed counter measures for the non-positive changes in a human body during space fight;
- Life support during space flight, including EVAs, and development of next generation biological systems for life support.

Together with the IBMP, a number of successful cooperative research projects between Roscosmos and the European space industry have been conducted. The Institute has deep experience in the ground simulation of long duration space flights with broad international cooperation. In November 2011, it completed the Russian project Mars-500, an imitation of human space flight to Mars. ESA, DLR, ASI, research establishments from Austria and Spain took part in it as well.

In 2017, the Institute began a series of joint Russian-American experiments with SIRIUS—the experiment involved the isolation of a group of people including one representative from Germany, Viktor Fetter, simulating the 17-days long trip to the Moon. According to Oleg Orlov, the director of IBMP, it is the beginning of the DSG project that will be continued in 2019–2020 (Pesljak, 2017).

4.2.2.2 Launchers and Launch Services

ESA and Roscosmos have established strategic and mutually beneficial relations in the area of launcher exploitation, again a strong asset of Russia's space activities. In this area, cooperation has been taking place with respect to the commercialisation of Russian launchers initially through Euro-Russian joint ventures such as Eurockot and Starsem.

- Eurockot was created in 1995 to commercialize the Rockot launch system for operators of LEO satellites.[7] It is a joint venture of EADS Astrium (now Airbus SD) and Khrunichev State Space Research and Production Space Center (KhSC), holding 51 and 49% respectively. It has dedicated launch facilities in Plesetsk Cosmodrome in Northern Russia. As of January 2018, Rockot is set to be retired following the two remaining launches.
- Starsem was established in 1996 to commercialise launches with Soyuz. Starsem's shareholders were EADS-Astrium (35%), Roscosmos (25%), Samara Space Centre (25%) and Arianespace (15%). The company was integrated into Arianespace in 2013 and is now within the Ariane Group.

Russia and Europe have also cooperated on the development of new launchers. R&T exchanges in this area began in 1991 with liquid-propulsion engines and, at

---

[7]Developed from refurbished intercontinental ballistic missiles (ICBM) components by Khrunichev, the Rockot is a three-stage launch vehicle predominantly used for sending scientific, Earth observation and climate monitoring satellites into LEO orbit. It can deliver up to 2150 kg in LEO from the Plesetsk launch centre and it launched EU Copernicus satellites Sentinel 3A, 2B and 5P in 2016.

the beginning of the 2000s, were extended to preliminary studies of future launchers, particularly reusable launch vehicles. More specifically, Russia (together with Ukraine) has been involved in the development of a third stage booster for the Vega launcher and has cooperated with CNES within the framework of the Oural programme for technology research on "medium- and heavy-lift launchers to succeed Ariane 5, offering the capacity to reach LEO, GTO, GEO, MEO and SSO, employing expendable, partly reusable (first stage or booster) and reusable technologies" (Centre National d'Etudes Spatiales, 2005).[8]

In addition, in 2003 ESA and Russia concluded an agreement for the exploitation of the Russian Soyuz launcher from the European Space Port in French Guyana. The European version of the Russian-manufactured Soyuz was introduced to consolidate Europe's access to space for medium-size missions, in particular to launch satellites up to 3200 kg into GTO and MEO, including constellations of two or more satellites. Named Soyuz-ST, it joined the family of European launchers in 2005, following the signature of a cooperation agreement between ESA and Roscosmos, which enabled Soyuz to use Europe's spaceport in Kourou. Needless to say, the French Space Agency CNES also played a crucial role in making such cooperation scheme possible, in particular by providing a strong backing with respect to the funding of the launch pad and addressing the security-related issues. The decision to develop a dedicated launch infrastructure at the Guiana Space Centre (GSC) was of interest to both Europe and Russia, as it enabled the former to complement the performance of the ESA launchers Ariane and Vega, and the latter to benefit from improved access to commercial markets and improved performance of the Soyuz launcher when being launched much closer to the Equator (Aliberti & Tugnoli, 2016). At that time, it was also in the interests of both Europe and Russia to boost their relationship in general, and cooperation in the launcher sector was seen as a tool for pursuing this objective (Fig. 4.7).

Soyuz-ST was launched from Kourou for the first time on 20 October 2011 and is designed to be used for medium-weight communication satellites as well as navigation and earth observation missions. The fact that the GSC can operate Soyuz has made it easier to use it for dual-use missions, as happened with the launch of France's Pleiades and ELISA satellites as well as the EU's Galileo constellation (Veclani & Darnis, 2014). From the very beginning, however, it has been questioned whether

---

[8] The OURAL programme, launched by CNES at the beginnings of 2000s but ultimately intended as a European undertaking, aimed "to design and build technology demonstrators for the development of a future launch vehicle in partnership with Russia" The agreement between CNES President Yannick d'Escatha and the Head of Roscosmos Anatoly Perminov covering cooperation on future launch vehicles and human spaceflight was signed on 15 March 2005 inn Paris. As reported by CNES, the ultimate aim of this launcher partnership programme was "to develop, by 2020, a new system combining cost effectiveness, a high level of reliability, enhanced safety and reduced environmental impact, to launch, transport and maintain satellites, spacecraft and orbital stations. This partnership with Russia, to be pursued in particular through joint construction of demonstrators, will seek to leverage the experience of Russia's launch industry and lay the groundwork for a possible future joint development, by bringing together the engineering cultures of Russian and European firms. The key technologies underlying these developments will involve a lot of study and experimental work, enabling European teams to maintain their expertise (Ibid).

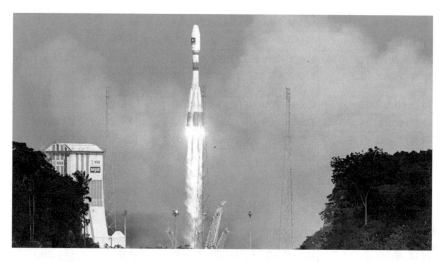

Fig. 4.7 Soyuz Launch from GSC. *Credit* Sputnik News

Europe should rely on Soyuz for missions having a high political-strategic value, considering that Soyuz is not a truly European launcher (Aliberti & Tugnoli, 2016).

As of 2018, the Soyuz ST launcher continues to be commercialised by Arianespace, though this commercialisation may be discontinued following the introduction of Ariane 6.2 in 2020 (see next section) (Table 4.5).

4.2.2.3 Space Sciences and Exploration

ESA and Russia have nurtured longstanding relations also in the area of space sciences. The most notable missions are:

- Rosetta
- Cluster
- Mars Express and Venus Express
- Integral.

The Russian Academy of Sciences is also providing an instrument for the ESA-led Bepi Colombo mission to "study and understand the composition, geophysics, atmosphere, magnetosphere and history of Mercury" (European Space Agency, 2017b). The mission is slated to launch in 2018 and the IKI of the Russian Academy of Sciences is to provide a Mercury Gamma ray and Neutron Spectrometer (MGNS).

Robotic exploration is emerging as the second pillar, after launchers, in ESA's Russia cooperation portfolio. The basis for such cooperation is the ESA-Roscosmos "Intergovernmental Agreement concerning Cooperation in Robotic Exploration of Mars and Other Bodies in the Solar System", which was negotiated in 2012 and signed in March 2013. More specifically, the agreement covers:

Table 4.5 The list of the launches from CSG including planned missions

Date	Payload	Mass (kg)	Orbit	Result
21 October 2011	Galileo IOV-1/2	1,580	MEO	Success
16 December 2011	Pleiades 1, SSOT, 4 x ELISA	2,191	SSO	Success
12 October 2012	Galileo IOV-3/4	1,580	MEO	Success
01 December 2012	Pléiades 1B	1,070	SSO	Success
25 June 2013	O3b F1	3,204	MEO	Success
19 December 2013	Gaia	2,105	L_2	Success
03 April 2014	Sentinel-1A	2,272	SSO	Success
10 July 2014	O3b F2	3,204	MEO	Success
22 August 2014	Galileo FOC FM1	1,607	MEO	Partial failure
18 December 2014	O3b F3	3,184	MEO	Success
27 March 2015	Galileo FOC FM3/FM4	1,597	MEO	Success
11 September 2015	Galileo FOC FM5/FM6	1,601	MEO	Success
17 December 2015	Galileo FOC FM8/FM9	1,603	MEO	Success
25 April 2016	Sentinel-1B, MICROSCOPE, 3 CubeSats	3,099	SSO	Success
24 May 2016	Galileo FOC FM10/FM11	1,599	MEO	Success
27 January 2017	Hispasat 36W-1	3,200	GTO	Success
18 May 2017	SES-15	2,302	GTO	Success
2018	MetOp-C		SSO	Planned
2018	O3b × 4		MEO	Planned
2018	OneWeb × 10		LEO	Planned
Late 2018	CHEOPS, COSMO-SkyMed		SSO	Planned

- **Robotic exploration of Mars**, first through the implementation of two ExoMars missions, which will provide the basis for subsequent stages of cooperation in this area;
- **Exploration of the Jovian system**, with particular focus on Jupiter and the Galilean moons, through the implementation of the ESA JUpiter ICy moons Explorer (JUICE) mission and the Roscosmos LaPlace-P Ganymede lander mission;
- **Exploration of the Moon**, with the long-term objective of delivering soil samples from the lunar polar regions, through the implementation of the Roscosmos Lunar Resource Lander and of the Lunar Polar Sample Return (LPSR) missions, with contributions to Roscosmos from ESA.

Table 4.6 ESA-Roscosmos ExoMars cooperation

	ESA contributions	Roscosmos contributions
2016 Mission	Trace gas orbiter Landing demonstrator module Scientific instruments	Launcher Scientific instruments
2020 Mission	Carrier spacecraft Rover Scientific instruments	Launcher descent module surface platform Scientific instruments

Between 2013 and 2017, the cooperation focus has been mainly on Mars exploration, starting with the ExoMars astrobiology programme, which over the past 15 years has undergone multiple phases of mission planning and international cooperation schemes.[9] The primary goal of the ExoMars programme is to search for evidence of life on Mars. The programme comprises two missions. The first, launched on 14 March 2016, consists of the Trace Gas Orbiter (TGO) and an Entry, Descent and landing demonstrator Module (EDM) named Schiaparelli. The second, planned for launch in 2020, consists of a rover and surface science platform (see Fig. 4.8).

As detailed by ESA, "TGO's main objectives are to search for evidence of methane and other trace atmospheric gases that could be signatures of active biological or geological processes. Schiaparelli has tested key technologies in preparation for ESA's contribution to subsequent missions to Mars. The 2020 rover will carry a drill and a suite of instruments dedicated to exobiology and geochemistry research. The 2016 TGO will act as a relay for the 2020 mission" (European Space Agency, 2017a).

The respective programmatic responsibilities of ESA and Roscosmos for the 2016 and 2020 missions have been detailed in the 2013 agreement and are summarised in Table 4.6.

Unlike the 2016 mission and other robotic exploration missions featuring international cooperation, an important feature of the 2020 ExoMars mission is that it is not limited to separate contributions (e.g. provision of scientific instruments or launch services) but is highly integrated with regard to all mission elements. Figure 4.9 provides a detailed indication of this integration (Fig. 4.9).

---

[9]ExoMars was conceived at the beginning of the 2000s as part of the broader ESA Aurora programme for the human Mars exploration of Mars. That initial plan envisaged the launch of a rover in 2009 and a sample return mission in subsequent stage. The programme was approved in December 2005 as an ESA optional programme, with the rover slated to launch in 2011 aboard a Soyuz launch vehicle. In July 2009, however, ESA and NASA signed an agreement (the Mars Exploration Joint Initiative) to join resources in the exploration of Mars. The agreement foresaw the utilisation an Atlas rocket launcher instead of a Soyuz, thus significantly altering the technical and financial setting of the ExoMars mission (including the combination of the mission with a second rover—the MAX-C—and a substantial reduction of the rover's weight to fit on the Atlas rocket). This notwithstanding, in August 2009 ESA signed an agreement with Roscosmos to secure a Proton launch vehicle as a "backup rocket" for the ExoMars rover, (which would include Russian-made parts) and to cooperate on the Phobos-Grunt programme. Owing to a budgeting reduction for NASA, in February 2012 NASA terminated its participation in ExoMars, forcing ESA to restructure the ExoMars programme. Eventually, on 14 March 2013, ESA and Roscosmos signed a new deal making Russia the full partner of ExoMars.

Fig. 4.8 Artist's view of
ExoMars. *Credit* ESA

4.2.2.4   Operational Satellite Systems

Another major area of Europe-Russia space cooperation is related to satellite systems, particularly in the field of electric propulsion and telecommunications. Concerning the former, since the 1990s Snecma (belonging today to the Safran Group) teamed up with the Russian manufacturer of plasma thrusters, EDB Fakel. Safran has been marketing the Fakel SPT100 stationary plasma thruster in Europe, through the joint venture ISTI, created with the American satellite manufacturer Space Systems/Loral. About 50 of these thrusters have been sold to Thales Alenia Space (TAS) and Astrium (now Airbus Space & Defence), and some of them have been put in service on three satellites: Intelsat 10, Inmarsat 4-F1 and Inmarsat 4-F2. "Safran has also been developing its own plasma thruster in collaboration with Fakel, the PPS 1350, based on European technologies. The first application of this thruster was on the ESA mission Smart-1, which it propelled into orbit around the Moon" (Mathieu, 2008).

Cooperation on telecommunications satellites has been taking place directly between companies. European companies have typically provided their Russian counterpart with payloads and systems for the platform of satellites intended for

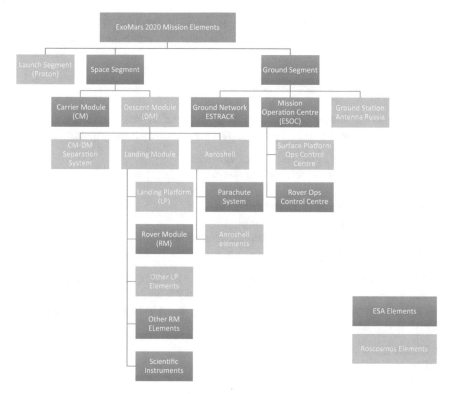

Fig. 4.9 ExoMars 2020 mission elements

the Russian market, to be launched on Russian launchers (Mathieu, 2008). Airbus SD has been mainly working with Khrunichev for over 20 years, while TAS has been working with NPO PM and with Khrunichev in this field. More recently, TAS won a bid to build the Yamal-601 telecommunications satellite, which is slated to launch in 2018 (Zak, Communication Satellites, 2017c).

Although in the past NPO PM also won an order from Eutelsat to build the Siberian-European Satellite (or Sesat), which became the first communications spacecraft built by a Russian prime contractor for a Western customer, cooperation with Europe in the field of telecommunication satellites illustrates the fact that Russia still lacks state-of-the-art technologies for satellites and, in order to get them, is buying them abroad and developing cooperation with European partners.

Cooperation is more particularly developed in the field of telecommunications than in other fields such as remote sensing where export restrictions hinder cooperation with some European countries (Mathieu, 2008). Indeed, in stark contrast to ESA cooperation with other countries (e.g. China), one area where ESA and Roscosmos have not come to terms is EO and related applications. This lack of cooperation in the field can be mainly attributed to a mismatch in capabilities as well as in core interests, even though Roscosmos has become part of the "International Charter on Space and

Major Disaster", and ROSHYDROMET, as already pointed out, is cooperating with EUMETSAT in the field of meteorology, oceanography and climate change.

4.2.2.5 Current Cooperation Assessment

An assessment of Europe's experience in its cooperation with Russia reveals that it has accrued a number of benefits for the two partners. The most important include:

- Complementarities in the technological systems
- Improving access to market and competitiveness (e.g. launchers)
- Cost sharing
- Increased efficiency and flexibility
- Increased stability of the programme
- Improved performance of systems and mission results
- Experience sharing.

   While Europe and Russia have a long cooperation history that has benefited them both, barriers to cooperation between Europe and Russia are still high. Building on the analysis of past cooperation experience, and interviews with key stakeholders, the most noticeable barriers are:

- Different standards and legal framework
- Customs and security regulations
- Language and cultural differences
- Different project processes (e.g. review cycles) and decision making
- Different schedules and priorities causing delays.

4.3 Evaluating the Impact of Russia's Evolution on Cooperation Prospects for Europe

When assessing the future evolution of Europe-Russia space relations, it is first essential to identify the impact that the recent evolutions of the Russian space programme, industry and international posture may have on Europe's need/will to cooperate with Russian partners. This will be used to determine foreseeable developments in key issue areas and a set of possible options for Europe.

4.3.1 Consequences of Russia's Space Sector Evolution

The profound transformations the Russian space programme has been undergoing over the past few years create Janus-faced implications for Europe-Russia space cooperation, simultaneously providing greater incentives for cooperation and posing some undeniable strains on it.

4.3.1.1 Institutional Consolidation

The impact of the 2015/16 institutional consolidation of Roscosmos on cooperation with ESA is minimal. Although the different legal status of Roscosmos SC may require some amendments to the cooperation agreements and cause some delays, the modus operandi of Russian enterprises and centres now merged into Roscosmos has not been altered. In the medium term, the highly hierarchical business approach could, on the contrary, streamline the implementation of cooperative agreements. In addition, the business mind-set brought by the new Roscosmos leadership and the fact that the current head Komorov does not come from the military environment—unlike previous heads of Roscosmsos—certainly bodes well for future cooperation with Europe.

4.3.1.2 Budget Reductions

As detailed in Chap. 2.2, the planned budget of 1406 billion RUB (€18.5 billion when approved) for FSP 2016–2025 is significantly less than the envisaged figure in the first drafts of the FSP back in 2013. The impact of FSP-2025 budget reductions on the various domains of space activities has not been formalised, though it is clear that some of the main FSP-2025 goals will be challenging to achieve without a budgetary increase—e.g. a new launcher for crew transportation, a new crew vehicle with the capability to fly to the Moon, and new launch infrastructure located at the Vostochny cosmodrome. While it remains to be seen in what direction the FSP-2025 budget profile will be amended based on the Russian economic situation, the additional reductions imposed by the Ministry of Finance for the 2017, 2018 and 2019 budgets have already generated short-term impacts (one of them is the reduction in the Russian ISS crew from 3 to 2).

Roscosmos has informed ESA that the joint cooperation projects, ExoMars and lunar exploration, have not been affected. According to Roscosmos, even in its reduced form, the FSP-2025 contains sufficient funding to meet Russian obligations under the joint ExoMars programme and under its future lunar robotics missions.

It should be noted that budget reductions not only leave planned cooperation unaffected but, to the contrary, may also induce Roscosmos to seek greater engagement with foreign players, particularly for under-funded budget items such as space sciences. Cooperation with ESA would not only aggregate resources for larger projects but also increase the effectiveness of its national programme, by freeing up resources and allowing Roscosmos to match its resources more effectively with its plans.

4.3.1.3 Policy Re-orientation

An analogous line of reasoning applies to the policy re-orientation enshrined in the FSP-2025, which clearly highlighted the effort by the current Roscosmos leadership to steer the industry toward more pragmatic goals rather than prestige-oriented projects inherited from the Soviet period. On the one hand, this new prioritisation decreases the scope of potential undertakings to be pursued through cooperation. On the other, it clearly compels Roscosmos to expand cooperative ties in order to retain/develop state-of-the-art technologies in key areas that it may otherwise lose. The Joint Statement on the DSG is a cogent example of such necessity.

The re-orientation in the programmatic objectives of the FSP-2025 opens valuable opportunities also for European stakeholders. More specifically, the change in the ranking of the fields of space activities with a clear priority for Earth observation/telecommunications (including enhanced exploitation of satellite applications) creates a clear convergence with ESA/EU space programmes, providing concrete incentives to cooperate in the field, which in fact is one of the cooperation areas identified by the Roadmap on the Common Economic Space. These incentives also stem from the current Russian eagerness to enhance exploitation not only on the military side, but also on the civilian and commercial ones.

In addition, the selection of the Moon as the priority destination for future exploration perfectly matches with ESA exploration strategy. Not surprisingly, ESA's participation in Roscosmos-led lunar exploration missions was included among the main programme activities covered by the Declaration on the European Exploration Envelope Programme (E3P), agreed and subscribed to at the 2016 Ministerial Council in Lucerne and now listed in a new Cooperation Agreement between ESA and Roscosmos.

4.3.1.4 Reliability Issues

The numerous lapses in quality control and reliability issues that have emerged over the past 10 years are certainly contributing to making the Russian partner less attractive for cooperative undertakings. With the string of embarrassing launch failures continuing in 2017, and insurance prices for Russian systems skyrocketing, the credibility of the Russian space industry is clearly endangered. This is certainly true for established players such as Europe or the U.S., but also for emerging space faring nations. As already reported in a previous ESPI study, Russian credibility is on the descent and "this is well understood by Russia's partners and possible partners. Ironically, this might give Russia some competitive advantages, alongside the obvious negatives. Russia is possibly seen as a less threatening partner than China, and [most of] Russian technology remains state-of-the-art, meaning that the aim of Russian partners might be to obtain licensing deals allowing domestic production of Russian technology, rather than relying on engineering implementation by the questionable Russian industry. This is hardly ideal for Russia, but perhaps one of the ways forward. Unfortunately, it is not a way forward for those relying on Russian launchers, however, because the recreation of launcher production lines is hardly feasible" (Hulsroj, 2014), as is evident from the failed production of RD-180 launcher engine technology in the U.S.

4.3.2 Consequences of Russia's International Posture Evolution

Like its domestic evolution, the new developments in Russia's external posture have the potential to impact European-Russian relations and space cooperation.

4.3.2.1 Russia's more Assertive Foreign Policy

The Russian leadership has adopted a more assertive—not to say aggressive—posture on the international scene, as evidenced by the utilisation of its newly upgraded military to carry out a series of threatening manoeuvres along European borders (in addition to Ukraine) as well as the recourse to cyber-attacks, disinformation campaigns and even interference in the political processes of several Western countries (Polyakova et al., 2017). This willingness to defend and promote national interests in a more assertive—not to say aggressive—fashion has alarmed many European capitals and inevitably led to an unfavourable political context in which to develop space cooperation.

Indeed, whereas for existing cooperation projects key European stakeholders such as ESA and EUMETSAT have been opting for a "business as usual" approach by keeping politics out of their activities, it is undeniable that there have been fewer drivers for to further expansion of space cooperation with Russia. Quite to the contrary, the strains in the political sphere—coupled with the loss of confidence in Russian space systems—seem to have made Europeans more hesitant in excessively relying on the Russian partner in key domains such as access to space.

This is for instance reflected in the decisions endorsed at the 2014 ESA Ministerial Council concerning the development of Ariane 6 and the possible discontinuation of Soyuz from Kourou. While that decision predominantly stemmed from commercial and industrial drivers, the development of a double configuration of Ariane 6 in fact responded to political considerations as well, namely the perceived need to overcome the reliance of European institutional launches on the availability and prices of Soyuz. Arguably, the tense political context that followed Russia's annexation of Crimea must have borne some weight on European decision-makers.

4.3.2.2 Sanctions on Russia

The economic sanctions imposed by Europe and the U.S. on Russia in the aftermath of the Ukraine crisis have not, as already mentioned, halted existing cooperation undertakings. However, as many interviewees have pointed out, they have made daily cooperation activities more cumbersome and caused some delays (e.g. the sanctions on electronic components have caused non-negligible difficulties in the execution of the ExoMars programme because a waiver on the purchase of hydrazine for the 2016 mission had to be issued).

In addition, economic sanctions have enhanced Russia's drive towards greater autonomy from foreign sources. This tendency intensified pursuant the Crimea crisis of 2014, when Russia decided to put in place an import substitution programme to develop production capabilities for critical components that were not yet domestically available, and to increase its technological non-dependence. In addition, to counter the immediate shortfalls, Russia also started to purchase electronic components from China (with a deal estimated worth several billion US$ in early 2015). Inevitably, this new development is bound to negatively affect the business of several European companies.

4.3.2.3 Russia Partnership Diversification

In parallel with its recent strive towards technological non-dependence, Moscow has been trying to reinforce its relations with other partners, particularly emerging space-faring nations and non-Western countries, in the form of both technology transfer and joint activities. As highlighted in Chap. 3, such diversification in the partnership portfolio is not necessarily detrimental for Europe, as Moscow has not shown less interest in Russia-Europe cooperation while re-orienting its space posture.

As already noted in a previous ESPI study (Mathieu, 2008), Russia's cooperative ties may be either complementary or substitutive to existing or future ones with Europe. Some of Russia's partnerships are complementary to those with Europe and beneficial to Europe an stakeholders. These include the activities within the ISS framework or scientific missions with other partners (e.g. Venera-D programme currently under consideration with the U.S.).

Other partnerships are dissociated from those with Europe, but have the potential to impact Europe's market and attractiveness. For instance, Russia's cooperation with African countries, such as South Africa or Angola, are examples of cooperation that may limit the interest of those countries in working with Europe (Ibid). Furthermore, the establishment of cooperative ventures within the framework of the BRICS may also reduce cooperation opportunities between ESA and these countries, including Russia itself.

While for Russia, Europe will remain a valuable source of technology, capital and (quality) management know-how, as well as a market and an access point to other markets, in some areas, Russia today has potential alternatives to Europe. Discussions with other partners, such as China, Brazil and Iran, offer Russia more options and demonstrate to Europe that Roscosmos has other possibilities as well. This is the case in access to space, for instance, where joint work is being discussed with China and Kazakhstan, and in the area of electronics, which could be provided to Russia by the Republic of Korea, China or India.

This diversification, however, should not be overstated either. As the failed cooperation initiatives with India in the field of launchers, space exploration and human spaceflight clearly show, not all desirable partnerships might actually materialise, and in some instances Russia's position in the international space arena may even border isolation. Indeed, Russia's present international position gives Europeans some bargaining power in shaping cooperation opportunities with Roscosmos. One can therefore expect cooperation in some fields such as space science and exploration, and telecommunication satellites, to be quite stable over the medium term.

In addition, some of the most recent developments in Russia's external relations (such as for instance the envisaged participation in the LOP-G framework) offer additional avenues for pursuing possible cooperation. The question to address is how strong is the need and will to cooperate with the Russian partner?

4.3.3 Possible Developments and Future Opportunities

As emerges from the analysis above, despite some negative impacts and limitations, there is abundant scope and opportunities for European stakeholders to continue cooperating with Russia in the foreseeable future. Russia is still considered, together with the U.S and, more recently, China, as a priority partner for Europe in its space endeavours. The main focus of Europe's cooperation with Russia will be the implementation of the many cooperative undertakings already under way in the areas of access to space, human spaceflight and space sciences and exploration (Soyuz from the GSC, Bepi Colombo, the ExoMars programme and the Roscosmos-led Luna programme).

In addition, new cooperation opportunities, specifically in the field of space applications, space sciences and human exploration in the post-ISS exploration context, could also be explored in the coming years by the two partners, provided an appropriate level of political will on both sides. However, foreseeable developments indicate that the impact of the degraded political context should not be underestimated and that for now that Europe pathway will be one of selective engagement rather than the fully-fledge partnership or even joint strategy envisaged in the 2000s.

This becomes evident when analysing the four major areas of current and potential cooperation between Europe and Russia: access to space, space sciences and exploration, human spaceflight and satellite applications, where a purely programmatically-driven approach is emerging.

4.3.3.1 Access to Space

Cooperation in the exploitation of Soyuz from French Guiana will continue smoothly over the next few years, despite the threat by Roscosmos in 2016 to suspend supply of Soyuz-ST to Arianespace (RT News, 2016), stemming from the freezing of payments for the vehicle manufacture imposed by a French court pursuant to the lawsuit against Russia by shareholders of the Yukos oil company (300 million euros) (de Selding, 2016). In 2018 alone, four launches of Soyuz-ST from CSG are scheduled—two for ESA (MetOp-C and the CHEOPS space telescope) and two commercial missions.

However, with the decision endorsed at the ESA Ministerial Council of 2014, Europe has decided to strengthen sovereign autonomy in access to space by realizing an all-European family of rockets (Ariane 6.2, Ariane 6.6 and Vega C) and thereby overcoming the current dependence on the Russian-made Soyuz.

Indeed, in parallel to the commercial drivers, the decision to proceed with a double configuration for Ariane 6 responds to political considerations as well, particularly the perceived need to overcome the present reliance of European institutional launches on the availability and prices of Soyuz. As a medium-lift launcher, Ariane 62 will amply cover the payload segment and will thus offer the possibility of replacing the Russian-manufactured rocket with a truly European launcher. Indeed, ESA's commitment to its development has made plain that Soyuz-ST will be progressively phased out and replaced with Ariane 62 and Vega-C (Aliberti & Tugnoli, 2016).

In addition to ensuring autonomy in a strategic payload segment, the key "political driver" behind the development of Ariane 62 is to help sustain industrial activities and research capabilities in ESA Member States rather than in Russia.

A similar line of reasoning applies to the evolution of the Vega rocket. In fact, given that the current Vega launcher includes non-European elements (namely, the AVUM fourth stage), a major element of the new Vega Consolidation strategy would be to fully'Europeanise' the new European launchers. With regard to the Vega programme it is worth noting that a further upgraded version of Vega for the mid-2020s is currently being studied: the Vega-E (Evolution), which at present is designed to feature P120 as first stage, Zefiro 40 as second stage, Cryogenic upper stage (LM-10 MYRA) replacing both the Zefiro 9 and AVUM+. If this configuration is eventually adopted, Vega-E would become a three-stage launcher, with no optional fourth stage. In that way, all components of the rocket would be built inside Europe and autonomy from foreign sources would be ensured (Ibid).

All in all, Europe's decision overcome the dependence on Soyuz and develop a purely European line of products shows that cooperation with Russia is currently valued only if associated with material opportunities rather than the possibility to foster closer political ties. As Moscow visualises it, in such a relation, Russia has been considered a supplier, if not a subcontractor; arguably, an approach far from the original idea of building a long-term-oriented partnership aimed at satisfying the two sides' goals. Not surprisingly, whereas cooperation with Russia in this strategic domain is set to decline over the next 10 years, (also considering the discontinuation of the Rockot exploitation by the JV Eurokot) it remains likely—and of course desirable—that Arianespace will try to continue commercialising the Soyuz launcher from the Baikonur launch site.

Arianespace's arrangement with Starsem—which envisages the possibility of using the Baikonur/Plesetsk Cosmodrome for the launch of Soyuz-ST in case of necessity—has been of utmost importance for Arianespace, as evidenced by the successful conclusion of the Arianespace-OneWeb contract in summer 2015 (a contract to launch roughly 700 LEO satellites aboard the Russian Soyuz from the cosmodrome of Baikonur) (de Selding, 2015). Therefore, considering the future phase out from Kourou and replacement of Soyuz-ST with Ariane 62 and Vega-C, European stakeholders could indeed contemplate the possibility of concluding a new agreement to ensure the availability of Soyuz from Baikonur/Plesetsk also in the next decade. In the current context, the possibility may prove particularly challenging from political, security and legal standpoints, but also highly beneficial for Europe.

In addition, given the future discontinuation of the close cooperation pursued over the past 20 years in this strategic area, an alternative and symbolic pillar of European-Russian space cooperation should be identified. Since it is unlikely that future cooperative undertakings could directly involve access to space—also in view of the general trend of strengthening national autonomy and privatisation or re-nationalisation of the industry—the historical capacity of Europe to act as a bridge- and coalition-builder could be more firmly enshrined into the broader framework of a "European Vision for Space", in which other options are explored. Among them, space exploration could indeed provide an equally symbolic and effective framework for cooperation in the post-ISS period, resulting in a bold international endeavour endorsed at the very highest political level (Aliberti & Tugnoli, 2016).

4.3.3.2 Space Science and Exploration

Unlike launchers, space sciences and robotic exploration are gradually emerging as a second pillar, after human spaceflight, in ESA-Roscosmos strategic relationship.

In this area, joint work will continue on the development of the ExoMars 2020 mission with the conduct of tests at subsystem and system level and the subsequent activities related to the Assembly, Integration and Testing (AIT) in the course of 2018/2019.

In parallel with the cooperation on Mars exploration—and building on its experience—cooperation with Roscosmos will be extended to lunar robotics projects. In October 2017, ESA and Roscosmos finalised a cooperation agreement defining near-term and medium-term objectives for incremental cooperation on the exploration of the Moon, namely:

- in the near term, the objective is to access and explore the lunar surface through the Roscosmos-led "Luna-Glob" Lander, "Luna-Resurs-1" Orbiter, and "Luna-Resurs-1" Lander missions;
- in the medium term, the objective is to investigate the feasibility of returning regolith samples to Earth for further exploration of the lunar polar regions through

the joint ESA-Roscosmos "Luna-Grunt"/Lunar Polar Sample Return (LG/LPSR) mission.

The launch schedule and a description of the contributions of these missions are provided in Table 4.7.

The Moon is ESA's top priority destination for robotic exploration as well as the next likely destination for human exploration following the decommissioning of the ISS (Huffenbach, 2014). To this end, ESA is developing core exploration products based on previous investments as a way to prepare for future exploration missions, namely:

- PILOT (Precise Intelligent Landing using On-board Technology)
- SPECTRUM (Space Exploration Communication Technology for Robustness and Usability between Missions
- PROSPECT (Platform for Resource Observations and in Situ Prospecting in Support of Exploration, Commercial Exploitation and Transportation)

Cooperation with Russia is considered key to support the development of these core products and ESA's broader exploration strategy, as evident from Table 4.8, which benchmarks ESA's lines of technological development (PILOT, SPECTRUM and PROSPECT) with its contributions to Russian-led missions (European Space Agency, 2015).

By pursuing this cooperation with Roscosmos, Europe (but also Russia) will ensure the development of key technologies for the pursuit of more ambitious exploration missions; technologies that in turn will contribute to making ESA a so-called partner of choice in the international exploration architecture of the post-ISS context. Clearly, while cooperation in this field is mainly intended to fulfil programmatic needs rather than political purposes, it should nonetheless continue to be backed at the political level.

The full list of planned cooperative missions conducted by ESA and Roscosmos in the field of space sciences and robotic exploration is presented in the Table 4.9.

Table 4.7 ESA-Roscosmos future lunar missions

Mission	Launch date	Objectives
Luna-Glob lander	2019	Validating spacecraft soft landing capabilities and long-duration operations in the south polar region of the Moon and conducting scientific research of the lunar exosphere and lunar regolith
Luna-Resurs-1 orbiter	2021–2022	Performing remote research from lunar orbit and relaying scientific data and telemetry from the Luna-Resurs-1 Lander
Luna-Resurs-1 lander	2022–2023	Examining the surface regolith and exosphere at high latitudes in the South Polar region and validating advanced lunar exploration capabilities
Luna-Grunt/LPSR	TBD	Collecting samples of lunar regolith in the south polar region of the Moon and returning them on Earth

Table 4.8 Mapping of ESA contributions to Russian missions

		Luna Glob lander	Luna-Resurs orbiter	Luna-Resurs lander	LPSR
SPECTRUM	Ground support	•	•	•	•
	Inter-spacecraft link		•	•	•
PILOT	Landing site characterisation			•	•
	Navigation for precision landing			•	•
	Hazard avoidance			•	•
PROSPECT	Drilling and sampling			•	•
	Sample processing and analysis			•	•

Source European Space Agency (2015)

Table 4.9 Future ESA-Roscosmos space sciences and exploration missions

Date	Project name	Comments
2018	BepiColombo	Mercury comprehensive study
2019	Luna-Globe (Luna-25)	Simple lunar lander
2020	ExoMars	Rover and surface science platform
2021	Luna-Resurs-1 (Luna-26)	Lunar Orbiter
2022	Jupiter Icy Moon Explorer	possible scientific exchanges with Ganymede mission
	Luna-Resurs-PA (Luna-27)	More advanced Lunar lander
2024	Phobos-Grunt-2	Phobos soil return mission
2025	Bion-M3	Biological-research satellite
2025+	Luna-Grunt	Lunar soil return mission

4.3.3.3　Human Spaceflight

In the field of manned spaceflight, ESA and Roscosmos will continue to cooperate within the framework of the ISS, the operations of which have been extended until at least 2024.

Europe will continue to rely only on Russia for crew access to the ISS, and training of its astronauts will also continue to be partially carried out in Star City, 50 km from Moscow. Following the successful missions of ESA astronaut Thomas Pesquet in 2016/17, and of ESA astronaut Paolo Nespoli in 2017, in 2018, Alexander Gerst will launch aboard Soyuz launcher to the ISS in May, when he will become the ISS commander and he will conduct a plethora of scientific experiments for ESA. Finally, ESA astronaut Luca Parmitano—who is assigned for a long-duration mission in 2019 aboard a Soyuz launcher—will conduct his training activities in Star City.

In terms of ISS utilisation, over the few next years the ESA Columbus laboratory will host the conduct of a joint Roscosmos-ESA Plasma-Kristall-4 experiment, together with other experiments on the adaptation of the human body to the conditions of short and long-term space flights in connection to the flight of ESA astronaut Luca Parmitano.

In addition, the ESA-developed European Robotic Arm (ERA) will be launched on the Russian Multi-Purpose Laboratory Module (MLM). As detailed by ESA, the ERA "will work with the new Russian airlock, to transfer small payloads directly from inside to outside the International Space Station. This will reduce the set-up time for astronauts on a spacewalk and allow ERA to work alongside astronauts" (European Space Agency, 2014). As graphically captured in Fig. 4.10, the ERA consists of "two end-effectors, two wrists, two limbs and one elbow joint together with electronics and cameras. Both ends act as either a 'hand' for the robot, or the base from which it can operate" (Ibid). Although no definitive schedule has been issued yet by Roscosmos, the launch is expected at the end of 2018.

The deployment of the ERA with the MLM and subsequent Russian elements for the ISS (see Chap. 2.2) will certainly ensure close cooperation activities between Europe and Russia for the years to come, but the outstanding issue for manned spaceflight cooperation remains the post-ISS context.

The destiny of the biggest modern space station must be decided in 2020. Three main options are possible:

- ISS operations will be prolonged till 2030;
- Roscosmos will create a smaller station based on existing (but yet to be launched) modules;
- Parties will decommission the ISS and work on the NASA-led LOP-G and lunar base together.

All these possible options imply different amounts of cooperation between ESA and Roscosmos. The current workflow is not likely to be changed until at least 2024. However, the preparation of the post-ISS era is "a central topic for the partners who, despite multiple exchanges of ideas, declarations and precursor programmes, have been, so far, struggling to build a steady, robust and commonly-shared vision for the future of human spaceflight and space exploration. Recent events and announcements suggest, however, that the state of affairs is now progressing as agencies seem to be converging toward a shared enthusiasm for the Moon and moving ahead with preliminary steps" (European Space Policy Institute, 2017). In this respect, the most promising scenario is given by the envisaged deployment by NASA of a Lunar Orbital Platform-Gateway (LOP-G) in cislunar orbit. The LOP-G concept—formerly known as Deep Space Gateway (DGS) "gained an official international dimension with the signature of a joint statement by Roscosmos and NASA on 27 September 2017 at the 68th International Astronautical Congress in Adelaide, Australia. Although NASA had already been discussing technical options for the DSG concept with ISS partners

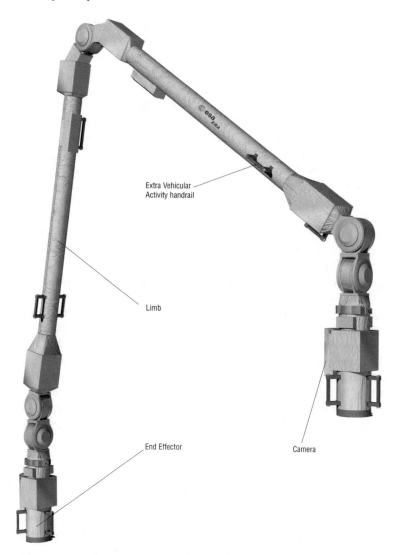

Fig. 4.10 The European robotic arm. *Credit* ESA

through the ISS Exploration Capabilities Study Team (IECST) and International Spacecraft Working Group (ISWG) during meetings in Japan, Canada and Europe in 2017, the joint statement, focusing at the moment on research studies, marks a noticeable milestone in the development of an international cooperation structure around the DSG project" (Ibid).

While it is important to recall at this stage that the LOP-G still has to face major hurdles before becoming a reality, if such cooperation eventually materialises, it will certainly provide Europe with a valuable framework to continue pursuing human spaceflight cooperation with its two historical partners, the U.S. and Russia. In addition, when looking at the rehabilitated Moon objective and taking into account that current LOP-G concept foresees both robotic and human Moon landings, "one cannot overlook the renewed light that is shed on ESA Director General, J. D. Woerner's Moon Village vision. Indeed, the current dynamic of the international space exploration scene could offer an interesting opportunity for Europe to implement, at least partially, this ambitious vision. With the declared objective to prepare a journey to Mars, the deployment of a Moon Base as a potentially European-led component of an International DSG programme would certainly offer a relevant test bed for the development and validation of key capabilities for future Mars mission such as In Situ Resource Utilization, Robotic-Human cooperative operations or ground base assembly among many others" (Ibid). Given that NASA has thus far declared to be not interested in taking the lead on a human lunar mission as a primary project, Roscosmos could potentially become a valuable partner for Europe in this respect, also considering that human lunar exploration was originally included among the future goals of the Federal Target Programme FSP-2025.

In general, however, the possibility of implementing one such cooperation framework and the role that Europe will hold within the next international partnership framework will, first and foremost, depend on the resolution of the still-pending political issues with Russia; second on the implementation of the LOP-G concept; third, on the prosecution of the planned cooperation between NASA, Roscosmos and ESA; and finally; on the financial and technical resources Europe is ready to commit. Once again, strong political support at the highest level is necessary to realise such a long-term vision, an ingredient that was, however, missing at the last ESA Ministerial Council, where a weak level of political backing was expressed by ESA Member States. How such political support may be found in Europe remains to date an open question.

4.3.3.4 Satellite Applications

Whilst satellite applications are an area where Europe and Russia have historically not come to terms, the recent re-orientation of the Russian space programme with the clear priority given to Earth observation, and to the enhancement in the exploitation of data and the development of applications for civilian, scientific and commercial purposes, potentially opens new opportunities for cooperative links.

Given the still sensitive nature of some areas of Earth observation, scientific applications, particularly in the field of Earth Sciences such as climatology, oceanography and meteorology, are optimal candidates to gradually encourage Europe-Russia cooperation in this field. A good basis already exists within this area (e.g. EUMETSAT cooperation with ROSHYDROMET) and there is a clear convergence of interests between Russian and European stakeholders (including the EU, ESA, EUMETSAT, as well as national space agencies) to strengthen activities in this field. As also highlighted by the EEAS, "the environment and climate change are areas of tremendous significance both to the EU and Russia. Given that long common land and sea borders, interconnected bio-systems, shared risks, and environmental problems are of common concern and should be addressed together, the need for joint action, together with the rest of international community, is even more pronounced in the area of climate change and global warming" (European External Action Service, 2017).

European and Russian commitment to contributing to a better understanding of climate change has been well highlighted by their long-standing policy dialogue on the topic (which unlike other policy dialogues has not been suspended pursuant to the Ukraine crisis, but is included within the policy of selective engagement pursued by the EU), as well as by their active support for the 21st Conference of the Parties (COP21). EUMETSAT has also shown interest in cooperating with Russia and other countries in the development of science-based climate services in synergies with its weather and ocean products.

As suggested for cooperation with other countries (Aliberti, 2018), this cooperation could be based ideally on a step-by-step approach. The first, functional step could be focused on satellite data and product exchange of respective Earth observation satellites in support of weather and climate monitoring. This should allow data access and redistribution rights to European satellite programmes by the Russian user community and vice versa. The overarching goal would not be just to gain access to more data, but rather to optimise the exploitation of existing capabilities and data through international synergies.

In addition, granted that exploiting the full potential of satellite data for climate products requires dedicated international efforts for recalibrating and reprocessing data, extracting climate records, and making them available to downstream applications and scientists—[10] such cooperation should be extended to also cover calibration/validation, data processing and re-processing methodology, and satellite data applications, including numerical weather and climate prediction. Towards this, and building on the experience already matured by EUMETSAT, an institutionally entrenched mechanism for cooperation could be established to deal with these tasks and to elaborate on joint research projects in climate research.

A next step in this cooperation would be the possibility of hosting payloads in respective Earth Observation missions (e.g. Russia's Electro and Meteor satellites) or flying a common set of instruments on their respective platforms so as to provide user communities with more homogeneous and consistent data flowing from similar technologies and better options for cross validation.

A final step in this cooperation trajectory would clearly be joint planning, development and operations of future satellite missions for climatology research. While ambitious, this cooperation format is already a reality in EUMETSAT-NOAA cooperation. Future cooperation with Russia—but also other players—could similarly follow this path (Aliberti, 2018). What is also important to highlight is that such

---

[10]Re-calibration and cross-calibration are an essential prerequisite for arriving at a homogenous time series of measurements across successive satellites that are useable for climate studies. A simple concatenation of data in time would not be useful for climate analysis. Therefore, in order to achieve an international, independent system for estimating global emissions based on internationally accepted data, cooperation to cross-calibrate instruments and cross-validate their measurements is quintessential.

cooperation would greatly contribute to ongoing international efforts (e.g. the World Meteorological Organization (WMO) Global Observing System, the Global Climate Observing System (GCOS), the United Nations Environmental Programme (UNEP), and other related programmes). Equally important, it would be a concrete milestone towards the possible development of climate services within the Global Framework for Climate Services (GFCS) established by the WMO. Finally, it could also enable additional cooperation formats in other areas of space applications, including commercial ones.

References

Aliberti, M. (2018). *India in space: Between utility and geopolitics*. Vienna: Springer.
Aliberti, M., & Tugnoli, M. (2016). *European launchers between commerce and geopolitics*. Vienna: European Space Policy Institute.
Centre National d'Etudes Spatiales. (2005, March 15). *Joint Press Release—CNES and Roskosmos Sign New Accord*. Retrieved January 3, 2018, from CNES: http://www.cnes-csg.fr/web/CNES-en/3269-joint-press-release-cnes-and-roskosmos-sign-new-accord.php.
de Selding, P. (2015, June 25). Launch Options were key to Arianespace's One-Web Win. *Space News*.
de Selding, P. (2016, October 25). *Roscosmos says Galileo, other European space programs could suffer from payment dispute*. Retrieved January 3, 2018, from Space News: http://spacenews.com/roscosmos-says-galileo-other-european-space-programs-could-suffer-from-payment-dispute/.
EUMETSAT. (2017). *Bilateral Cooperation*. Retrieved November 20, 2017, from EUMETSAT: https://www.eumetsat.int/website/home/AboutUs/InternationalCooperation/BilateralCooperation/index.html.
European Commission. (2003, January 24). Space Research: Joint Workshop to promote EU-Russia cooperation. *Press Release IP/03/118*.
European Commission. (2004, November 23). *EU/Russia: The four "common spaces"*. Retrieved December 11, 2017, from European Union: http://europa.eu/rapid/press-release_MEMO-04-268_en.pdf.
European Commission. (2015). *European economic forecast—Spring 2015*. Luxembourg: Publications Office of the European Union.
European Commission. (2017a, February 15). *Russia—Trade*. Retrieved 1 February, 2018, from European Commission: http://ec.europa.eu/trade/policy/countries-and-regions/countries/russia/.
European Commission. (2017b, November 11). *European Union, Trade with Russia*. Retrieved January 10, 2018, from European Commission: http://trade.ec.europa.eu/doclib/docs/2006/september/tradoc_113440.pdf.
European External Action Service. (2005, May 10). *Roadmap For The Common Economic Space*. Annex 1. Retrieved December 9, 2017, from European External Action Service: https://eeas.europa.eu/sites/eeas/files/roadmap_economic_en_2.pdf.
European External Action Service. (2017, November 21). *The European Union and the Russian Federation*. Retrieved December 3, 2017, from European External Action Service: https://eeas.europa.eu/delegations/russia/35939/european-union-and-russian-federation_en.
European Leadership Network. (2016, June). *What is the future for EU–Russia relations?* London: European Leadership Network.
European Space Agency. (2014, April 17). *European Robotic Arm*. Retrieved December 28, 2017, from ESA: https://www.esa.int/Our_Activities/Human_Spaceflight/International_Space_Station/European_Robotic_Arm.
European Space Agency. (2015). *ESA's plans for Lunar Exploration*. Noordwijk: ESA-ESTEC.

European Space Agency. (2017a, March 29). *What is ExoMars?* Retrieved December 1, 2017, from European Space Agency: http://www.esa.int/Our_Activities/Space_Science/ExoMars/What_is_ExoMars.

European Space Agency. (2017b, June 9). *BepiColombo Overview.* Retrieved November 28, 2017, from European Space Agency: http://www.esa.int/Our_Activities/Space_Science/BepiColombo_overview2.

European Space Policy Institute. (2017, November). *Next Steps to the Moon: What Role for Europe?* Vienna.

Facon, I., & Sourbès-Verger, I. (2007). La Coopération Spatiale Russie-Europe, une Entreprise Inachevée. *Géoéconomie 43*(1).

Fonseca, A. (2013, May 2013). *ESA cooperation with Russia, China, Brazil, India and South Africa.* Leiden: European Space Agency.

Huffenbach, B. (2014). Role of the Moon in ESA's Space Exploration Strategy. *NASA Community Workshop on the GER.* Washington D.C: ESA.

Hulsroj, P. (2014). The psychology and reality of the financial crisis in terms of space cooperation. In C. Al-Ekabi et al. (Eds.), *Yearbook on Space Policy 2011/2012. Space in times of financial crisis.* Springer: Vienna.

Mathieu, C. (2008). *Assessing Russia's space cooperation with China and India: Opportunities and challenges for Europe.* Vienna: European Space Policy Institute.

Pesljak, A. (2017, November 7). *TASS, A new experiment called SIRIUS starts in the Institute of Biomedical Problems.* Retrieved November 8, 2017, from TASS: http://tass.ru/kosmos/4706420.

Polyakova, A., Kounalakis, M., Klapsis, A., Germani, L. S., Iacoboni, J., Lasheras, F. D., et al. (2017). *The Kremlin's Trojan Horses 2.0. Russia's Influence in Greece, Italy and Spain.* Washington D.C.: Atlantic Council.

RT News. (2016, October 26). *Russia's Roscosmos threatens to halt rocket supplies to French company over frozen assets.* Retrieved January 28, 2018, from RT News: https://www.rt.com/news/364534-russia-suspend-arianespace-rocket/.

Russel, M. (2016). *The EU's Russia policy.* Brussels: European Parliamentary Research Service.

Veclani, A. C., & Darnis, J.-P. (2014). European space launch capabilities and prospects. In K.-U. Schrogl et al. (Eds.), *Handbook of space security. policies, applications and programs* (Vol. 2, pp. 783–800). New York: Springer.

VEDOMOSTI. (2016, January 13). *Цифры (Figures).* Retrieved December 16, 2017, from VEDOMOSTI: https://www.vedomosti.ru/newspaper/articles/2016/01/13/623928-tsifri.

Zak, A. (2017c). *Communication Satellites.* Retrieved December 10, 2017, from Russian Space Web: http://www.russianspaceweb.com/spacecraft_comsats.html.

Chapter 5
Conclusions

After feeding the impression of a second golden age for space activities during the course of the 2000s, over the past 10 years, the Russian space sector seems to have once again glided on the precipice of a systemic state of crisis. This situation has been the inevitable by-product of the persistence of deep-rooted problems (such as an outdated industrial base, gaps in quality control, excessive manufacturing capacity, low labour productivity, recurrent administrative reforms etc.) combined with a series of negative external factors outside the direct control of Russia's space authorities (in particular a tense political situation in the international arena and the still-ongoing financial and economic crisis).

The combination of these factors has caused a number of important consequences for the Russian space programme, the most prominent of which are:

- the unceasing sequence of launch and spacecraft failures occurred between 2010 and 2017,
- the delay—and in some cases the cancellation—of several programmes (e.g. the MLM, UM and NEM modules for human exploration, Venera-D, Venera-Glob, and Phobos-Grunt for planetary exploration, etc.),
- the substantial reduction in both short and long-term budget planning and disbursement,
- the strains on the country's commercial operations and cooperative undertakings.

Beyond the immediate impacts, the combination of these endogenous and exogenous factors has generated broader transformations in Russia's institutional space environment; in its policy orientations, and in its programmes and technological development. More specifically:

- At software level, there has been a fundamental re-orientation of Russia's space priorities towards more programmatic endeavours as a way to cope with the new budgetary realities;
- At orgware level, there has been an acceleration and change of directions in the institutional and industrial overhaul initiated in the mid-2000s.

This reform process, which was completed 2017, is certainly an indicator of the attention that the Russian government continues to give to national space activities,

© Springer International Publishing AG, part of Springer Nature 2019
M. Aliberti and K. Lisitsyna, *Russia's Posture in Space*, Studies in Space Policy 18,
https://doi.org/10.1007/978-3-319-90554-9_5

which are still considered a strategic asset for supporting the realisation of national objectives. However, it should not go unnoticed that in executing these reforms, Russia has consciously chosen not to follow the worldwide trends of strengthening the involvement of private players in the space sector, but rather to rely on a centralised public space sector (Roscosmos Space Corporation) and to focus on fostering its competitiveness in international markets, building on the successful model of the Rosatom state corporation. While the rationales are clear, the consequences of this strategic choice are still difficult to foresee. In fact, the assessment performed in this study pinpoints Janus-faced implications:

- From a positive angle, the recent centralisation may likely help in overseeing and addressing the quality issues that have arisen frequently in the past 10 years. In addition, the consolidation could have very positive impacts for the space sector by permitting streamlining and optimisation of both infrastructure and workforce, responding to the need for improved economic performance in the light of space budget cuts, while the structural change into a state corporation could serve to ease bureaucratic requirements.
- On a negative note, the recent changes in the orgware have made an increasingly blurred demarcation line between policymaking, hardware ordering and vendor functions. This point is even more important if one considers that one of the original goals of creating URSC was to separate client and policy aspects from the contractor sections. Without clear internal boundaries, a situation comparable to that existing before the entire reform process was started, similar to the functioning during the Soviet era, may be recreated. In addition, the change of status also means that Roscosmos is no longer obliged to publically disclose information on tenders for example, somewhat reducing overall transparency. Moreover, even though the lack of privatisation and support for private industry (with the partial exception of the Skolkovo space cluster) does not automatically preclude Russia's successful integration in global space markets, it does not seem particularly encouraging in the current setting either. This is particularly true in the global market for downstream service applications, where Russia's sales are in a tiny pocket, but the situation is not stable even when it comes to traditionally successful areas such as commercial launch services, in which Russia now sees its market share inexorably declining. More broadly, the lack of private players' involvement in the Russian space sector may decrease the prospects for technological innovation and the socio-economic impact of space-related investments, eventually causing overall stagnation of the Russian space programme.

As in the past, Russia will certainly continue to defy prediction about the eventual collapse of its space programme, but it appears now more than ever that in order to ensure the long-term competitiveness of its space sector, it will need to a devise a forward strategy and further open its space programme to international cooperation.

The analysis of Russia's relations with the major and emerging space faring nations has revealed a dichotomous evolution of its cooperation with foreign partners. On the one hand, while Russia has generally remained very open to international collaboration in space matters, the recent vicissitudes of its space sector and the strains in its economic performance and international relations (e.g. Ukraine, Crimea, Syria) have led to increasing tensions with Western countries (particular the U.S.), a sensible reduction of cooperation with some former partners (e.g. Ukraine) and, more broadly, to a strive towards greater autonomy from foreign sources. This tendency has intensified since the 2014 Ukrainian conflict, a year that marked the end of the cooperative undertakings that Russia put in place in the launcher sector in the mid-1990s. While unfastening the multinational joint-ventures (ILS, Sea Launch, ISC, SIS), and repatriating their shares to Russian entities, Russia put a stop to future use of vehicles not entirely manufactured domestically (Dnepr and Zenit) and accelerated the construction of the Vostochny launch site on its territory to replace its dependence on Baikonur. It more broadly started to develop production capabilities for critical components that were not yet domestically available.

On the other hand, in order to both compensate for the increasing strains on cooperation with the West, and to find alternative sources of revenues because of budget reductions, Moscow has been trying to reinforce its relations with other partners, particularly emerging spacefaring nations such as Brazil, South Korea and Indonesia, and non-Western countries, such as China and Iran, or associations such as the BRICS, in the form of both technology transfer and joint activities.

As emerges from the analysis, such diversification in Russia's partnership portfolio may cause important transformations in the international space alignments, but it should not be overstated either. Russia's past and current cooperation experience with countries such as India, China and Brazil shows that relations face inherent hurdles, including a mismatch in interests, priorities and capabilities, as well as a diffuse level of political mistrust. As the failed cooperation initiatives with India in the field of launchers, space exploration and human spaceflight clearly show, not all desirable partnerships might actually materialise, and in some instances Russia's position within the international space arena may even border on isolation.

In addition, some of the most recent developments in Russia's external relations (such as for instance the envisaged participation in the LOP-G framework and the recently established cooperation with ESA for lunar exploration) clearly show that Western countries remain preferred—not to say obliged—partners for Russia in its space endeavour. Russia's industry and research institutes are indeed interested in working more closely with their American and European counterparts (not least because Russia needs their quality management expertise). The Europeans in particular are perceived as important and reliable partners, with a diversified and attractive offer for Russia, and with their long-lasting cooperation experience having accrued tangible benefits for both sides.

All this might give Europe some competitive advantage when approaching the Russian partner, along with some negative drawbacks. However, the domestic and international evolution of the Russian space programme is proving to have partially hampered the will of Europe to cooperate with the Russian partner.

For one thing, the numerous lapses in quality control and reliability issues that have emerged over the past 10 years have certainly contributed to making Russia less attractive for cooperative undertakings. In addition, Russia's planning and budgeting no longer appear stable enough to reach long-term goals, a fact that may potentially endanger ongoing cooperative activities and necessitate that future cooperation areas be chosen wisely as to avoid delays or even cancellations. More broadly, the overall political context is not very favourable for developing cooperation in space or elsewhere. The assertiveness of President Putin has created apprehension, or at least concern, about the Russian leadership's intentions, and has predictably dwindled dialogue opportunities with Europe. Given the recent re-election of President Putin and the still difficult the relationship between Moscow and Washington, Europe undeniably finds itself in a rather uncomfortable position. Therefore, given the many mutual interests, a shift towards a more constructive attitude in the political interplay is necessary to expand cooperation.

Despite the political context and the possible drawbacks caused by the state of crisis in the Russian space industry, Europe and Russia remain today very interdependent in their space endeavours, and ESA continues to see plenty of opportunities to further mutually beneficial cooperation with Russia, as, inter alia, demonstrated by the new Cooperation Agreement of 2017. It should also come as no surprise that, in the midst of the Ukrainian crisis, the Resolution adopted at the ESA Council at Ministerial Level of 2014 explicitly acknowledged Russia as one of the three strategic partners of ESA, together with the U.S. and China.

Whereas Europe-Russia ties are certainly set to decline in one of their most important fields of cooperation, namely launchers and launch services, they are now further expanding in the area of space exploration with a number of ambitious projects throughout the next decade (ExoMars, Luna-Glob, Luna-Resurs, LPSR), and, thanks to the re-orientation of Russia's space policy, possibly also in the area of space applications for scientific and commercial purposes.

For the time being, the main focus of Europe's cooperation with Russia will continue to be the implementation of the many cooperative undertakings already under way in the areas of space sciences, human spaceflight and space exploration (Bepi Colombo, ISS-related activities, the ExoMars programme, etc.). A noteworthy evolution is now offered the cooperation activities surrounding the Roscosmos-led Luna programme (Luna-Glob, Luna-Resurs Orbiter, Luna-Resurs Lander, and Luna-Grunt sample return mission). The Moon is ESA's top priority destination for robotic exploration as well as the next likely destination for human exploration following the decommissioning of the ISS. By pursuing this cooperation, Europe (but also Russia) will ensure the development of key technologies for the pursuit of more ambitious exploration missions; technologies that in turn will contribute to making ESA a so-called partner of choice in the future international exploration architecture. Provided sufficient backing on both sides, this blossoming area of cooperation between ESA and Roscosmos may well replace the declining cooperation on access to space as a new symbolic—and strategic—pillar of cooperation, underpinning Europe's historical capacity to use space as a tool for bridge- and coalition-building, and possibly even pave the way for a more ground-breaking cooperation in the post-ISS human exploration context.

Annex A
Russia's Space Firsts

Year	Major achievements
1957	First intercontinental ballistic missile, the R-7 Semyorka First satellite, Sputnik 1 First living creature in space, the dog Laika on Sputnik 2
1959	First firing of a rocket in Earth orbit, first artificial satellite of the Sun, first telemetry to and from outer space, Luna 1 First probe to impact the moon, Luna 2 First pictures of the moon's far side, Luna 3
1960	First animals to safely return from Earth orbit, the dogs Belka and Strelka First probe launched to Mars, Marsnik 1
1961	First probe launched to Venus, Venera 1 First man in space Yuri Gagarin on Vostok 1
1963	First woman in space, Valentina Tereshkova
1965	First EVA, by Aleksei Leonov First probe to hit another planet (Venus), Venera 3
1966	First probe to make a soft landing on and transmit from the surface of the Moon, Luna 9 First probe in lunar orbit, Luna 10
1967	First unmanned rendezvous and docking
1969	First docking between two manned craft in Earth orbit and exchange of crews, Soyuz 4 and Soyuz 5
1971	First space station, Salyut 1 First probe to orbit another planet (Mars), first probe to reach surface of Mars, Mars 2
1975	First probe to orbit Venus, first photos from surface of Venus, Venera 9
1986	First permanently manned space station, Mir, which orbited the Earth from 1986 until 2001
1987	First crew to spend over one year in space, Vladimir Titov and Musa Manarov on board of TM-4—Mir

© Springer International Publishing AG, part of Springer Nature 2019
M. Aliberti and K. Lisitsyna, *Russia's Posture in Space*, Studies in Space Policy 18,
https://doi.org/10.1007/978-3-319-90554-9

Annex B
Legal and Policy Framework of Russian Space Activities

National Space Legislation

In addition to international legal instruments and the Constitution of the Russian Federation adopted in 1993, Russian space activities are regulated by several domestic legal instruments. There most relevant are:

Federal Laws and Codes, which include:

- On Space Activities (1993 and amendments)
- On Navigation (2009, and amendments)
- On Licensing Specific Activities (2011)
- Federal Codes (Civil, Custom, Administrative, Penal, Land, Air, Tax…).

Regulations by the President and the Government, including:

- On Space Activities of the Russian Federation in 2013–2020 (2012)
- On the Development Strategy of the Rocket and Space Industry through the year 2013 and beyond (2012)
- On the Russian Academy of Sciences (2013)
- On the Federal Space Programme of Russia for 2016–2025 (2016)
- On the Federal Target Programme "Maintenance, development and use of the GLONASS system" for 2012–2020 (2012)
- On the Federal Target Program "Development of Russian cosmodromes" for 2016–2020 (2016).

Regulations by the Federal Ministries and Agencies, which include:

- Roscosmos

  - Administrative Regulation on Licensing of space activities (2012)
  - Administrative Regulation on Registration of space objects (2010)
  - Administrative Regulation on Cosmonaut selection and training (2010)

© Springer International Publishing AG, part of Springer Nature 2019
M. Aliberti and K. Lisitsyna, *Russia's Posture in Space*, Studies in Space Policy 18,
https://doi.org/10.1007/978-3-319-90554-9

- Ministry of Economic Development

  - Order determining the geodetic and cadastre equipment for GLONASS or GLONASS/GPS means (2012)

- Ministry of Transport

  - Order on the equipment of transport vehicles for GLONASS or GLONASS/GPS means (2010)

International Treaties and Arrangements

At international level, the Russian Federation, as a successor state to the Soviet Union, ratified most of the main international treaties governing space activities and is part of a number of arrangements relevant to space.

Item	Participation
Outer space treaty	Ratified
Rescue agreement	Ratified
Liability convention	Ratified
Registration convention	Ratified
Moon agreement	–
Nuclear test ban	Ratified
ITU	Ratified
MTCR	Participant
Wassenaar agreement	Participant
Hague CoC ballistic missile	Participant
Charter space and major disasters	Participant

As a successor state of the Soviet Union, the Russian Federation has also membership to the main space-related organisations, as detailed in Russia's membership in the main space-related international organizations.

Item	Membership
Intersputnik	Member
Intercosmos	Member
IMSO/Inmarsat	Member
ITSO/INTELSAT	Member
Eutelsat IGO	Member
GEOSS	Member
KOSPAS-SARSAT	Member

In terms of bilateral and multilateral agreements, the Russian Federation has signed:

- 20+ framework intergovernmental agreements (in addition to its agreements with the CIS countries, Russia has intergovernmental agreements with: Australia, Belgium, Brazil, Bulgaria, Chile, China, ESA, EUMETSAT, France, Germany, Hungary, India, Indonesia, Italy, Japan, Malaysia, Mexico, Republic of Korea, South Africa, Spain, Thailand, United States);
- 50+ multilateral and bilateral agreements (including the multilateral intergovernmental agreements on the ISS, on launches from Baikonour, on Cospas-Sarsat, etc.)
- 30+ interagency agreements.

Annex C
Russia's Major Programmes Overview

The Rescheduling of Roscosmos' Major Programmes

Field	Mission	Inception	Original date	Current date
Space transportation	Angara 1.2	1994	2005	2017
	Angara 5	1994	2005	2014
	Angara 5P	2012	2018	Cancelled
	MRKS	2005	2016	2030+
Human spaceflight	MLM	1997	2007	2018?
	UM	2009	2013	2018?
	SPM	2009	2014	2019
	OKA T-MKS	2006	2012	2019
	Soyuz MS	2009	2012	2016
	PTK-NP	2009	2015	2021
Space sciences	Spektr-R	1982	1997	2011
	Spektr-RG	1990?	2000	2017
	Lomonosov	2006	2010	2016
	Bion-M	2006	2013	2021
	Gamma 400	2013	2018	2025+
	Vozvrat-MKA	2009	2016	2025+
Space exploration	Phobos-Grunt	1992	1998	2011
	Phobos-Grunt 2	2011	2016	2025+
	Luna-Glob	2006	2012	2019
	Luna-Resurs	2009	2018	2021
	Luna-Grunt	2006	2016	2024
	Mercury-P	2006	2016	2020?
	Venera-D	2010	2016	2026

(continued)

© Springer International Publishing AG, part of Springer Nature 2019
M. Aliberti and K. Lisitsyna, *Russia's Posture in Space*, Studies in Space Policy 18,
https://doi.org/10.1007/978-3-319-90554-9

(continued)

Field	Mission	Inception	Original date	Current date
Application satellites	Kondor	1993	1997	2013
	Tundra	1999	2005	2015
	Kanopus-ST	2000	2006	2015
	Kanopus-V	2006	2013	2017
	Resurs-P	2006	2010	2014
	Luch-M	2002	2008	2015
	Bars-M	2007	2012	2015
	Ozbor-O	2012	2015	2023

FSP 2025 Programmes' Overview

Field	Mission	Launch Date	Launcher	Launch Site
Space transportation	Soyuz MS	2016 × 2 2017 × 2 2018 × 2 2019 × 2 2020 × 2 2021 × 2 2022 × 2 2023 × 3 2024 × 4 2025 × 2	Soyuz FG/Soyuz 2.1a	Baikonur
	Progress MS	2016 × 3 2017 × 3 2018 × 3 2019 × 3 2020 × 3 2021 × 3 2022 × 3 2023 × 3 2024 × 3 2025 × 1	Soyuz-2.1a	Baikonur
	Federation	2020 × 1 2023 × 2	Angara A5P*	Vostochny
Human spaceflight—ISS	UM	2018 × 1	Soyuz-2.1b	Baikonur
	MLM	2018 × 1	Proton-M	Baikonur
	SPM	2019 × 1	Proton-M	Baikonur
	EM	2025 × 1	Soyuz-2.1b	Vostochny
	OKA-TMKS	2019 × 1	Soyuz-2.1b	Vostochny

(continued)

(continued)

Field	Mission	Launch Date	Launcher	Launch Site
Space sciences	Lomonosov	2016 × 1	Soyuz-2.1a	Vostochny
	Rezonans	2021 × 1	Soyuz-2.1a	Vostochny
	Spektr-RG	2017 x1	Zenit M	Baikonur
		2021 × 1	Proton M	Baikonur
	Bion M	2021 × 1 2025 × 1	Soyuz 2 1b	Baikonur
Space exploration	ExoMars	2016 × 1 2018 × 1	Proton-M	Baikonur
	Luna Glob	2019 × 1	Soyuz 2 1b	Vostochny
	Luna Resurs	2020 × 1 2021 × 1	Soyuz 2 1b	Vostochny
	Luna Grunt	2024 x1	Angara A5	Vostochny
	Expedition-M	2024 × 1	Angara A5	Vostochny
EO satellites	Meteor-M	2016 × 1 2017 × 1 2019 × 1 2021 × 2 2022 × 1 2024 × 1	Soyuz 2.1b	Baikonur / Vostochny
	Electro-L	2017 × 1 2019 × 1 2024 × 1	Proton M	Baikonur
	Electro M	2025 × 1	Angara A5	Vostochny
	Arctica-M	2017 × 1 2019 × 1 2020 × 1 2024 × 1 2025 × 1	Soyuz 2.1b	Baikonur
	Resurs-P	2016 × 1 2018 × 1 2019 × 1 2020 × 1 2021 × 1 2023 × 1 2024 × 1	Soyuz 2.1b	Baikonur / Vostochny
	Kanopus	2016 × 1 2017 × 2 2018 × 2	Soyuz 2.1a	Baikonur / Vostochny
	Kondor-FKA	2018 × 1 2021 x1	Soyuz 2.1a	Plesetsk
	Smotr	2018 × 1 2019 × 6	Soyuz 2.1b	Vostochny
	Obzor	2021 × 1 2023 × 2 2024 × 1 2025 × 1	Soyuz 2.1a	Vostochny

(continued)

(continued)

Field	Mission	Launch Date	Launcher	Launch Site
Communication satellites	Gonets-M	2016 × 3 2018 × 3 2019 × 3 2021 × 3 2022 × 3 2023 × 3 2024 × 3 2025 × 3	Angara 1.2 Rockot	Plesetsk
	Express	2018 × 2 2019 × 2 2020 × 1 2021 × 2 2022 × 2	Proton M	Baikonur
	Yamal	2018 × 1 2019 × 1 2021 × 1 2022 × 2 2023 × 1	Proton M	Baikonur
	Luch-5M	2019 × 1 2021 × 1 2022 × 1 2024 × 1	Angara A5	Vostochny

As of March 2018, some of the listed missions have undergone changes in their launch schedule or the launch vehicle that will be deployed. The most important is the replacement of the Angara 5P rocket with Soyuz 5 for the launch of PTK Federation, whose maiden flight may also be postponed.

FSP-2025 Planned Launch Activity Overview

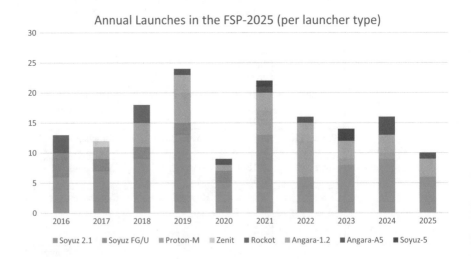

Annex D
The Evolution of International Joint Ventures

Sea Launch

The Sea Launch joint venture was created by NPO Energia, Boeing and the Norwegian company Kvaerner Maritime, with the subsequent addition of Ukrainian KB Yuzhnoe as the Zenit manufacturer. It was a joint venture for commercial launches (mostly of telecommunication satellites) of Zenit rockets from equatorial regions. The Sea Launch platform was one of the most dramatic ventures in space history. The idea of the sea-based launch site was pioneered by the U.S. Navy's Bureau of Ships around 1963 and then developed within the Soviet space programme. Maximizing the payload of the rocket by flying from the Equator, and having a movable launch facility, would offer great flexibility in terms of conceivable orbits for potential customers.

Unfortunately, this project faced an enormous number of problems, which included: underestimated costs; the merging of Boeing with McDonnell Douglas in 1996, which made it responsible for the Delta family's rockets that were Zenit's competitors; military secrecy inside the former USSR, and ultimately; the accident that took place in January 2007 when the rocket exploded during lift-off from the platform with New Skies' satellites on board.

Eventually the Sea Launch Company had to file for bankruptcy on 22 June 2009 and spent several years emerging from this state under the ownership of RKK Energia, until a new failure happened in 2013. Following this, RKK Energia had then asked the government to take over this troubled venture. At that time the company already owed its creditors $530 million. The head of RKK, Energia Vitaly Lopota, kept asking the government for help which included shifting some federal missions to Sea Launch. Meanwhile the original majority stakeholder, Boeing, decided to abandon the project and asked for the pay off of its $500-million share as part of its initial investments. Norwegians returned the money while RKK Energia and KB Yuznoe refused to do so. In August 2014 Sea Launch announced that it would put a hold on all operations until the middle of 2015 and laid off most of its personnel.

© Springer International Publishing AG, part of Springer Nature 2019
M. Aliberti and K. Lisitsyna, *Russia's Posture in Space*, Studies in Space Policy 18,
https://doi.org/10.1007/978-3-319-90554-9

RKK Energia lost the case against Boeing on 12 May 2016 and ended up with a debt of $322.49 million, while KB Yuznoe had to pay almost $193 million. Hope for the Sea Launch project started to re-emerge when the Russian airline company, the S7 Group, agreed to buy the assets of the project, including both an assembly and command ship and the launch support facilities in Long Beach, California, their home port, for $150 million. The deal was announced at the IAC 2016 in Guadalajara. S7 decided to leave the Sea Launch platform at Long Beach port and to continue using the Zenit rocket, and eventually received an agreement from Yuznoe.

Joining forces, Roscosmos and S7 managed to solve some problems including the huge debt, part of which was «substantially» covered by the deal with S7 according to President Vladimir Putin during his meeting with the head of Roscosmos, Igor Komarov, in 2016. The other part was unexpectedly returned to Boeing in the form of places for astronauts of U.S. choice for the human space flight program. This timing could not have been more fortuitous as Boeing was certainly late with its manned spacecraft Starliner and had no chance of fulfilling its obligations to the U.S. government.

Unfortunately, KB Yuznoe faced difficulties in obtaining certain components that had been manufactured in the Lugansk region, which was now cut off from the rest of Ukraine due to the military conflict between the government and pro-Russian rebel forces. The solution was rather unexpected: on 8 October 2017 Vladimir Solntsev from RKK Energia announced that Russia would continue to manufacture the RD-171 engines, while the other components would be made by the Ukrainian side but, due to the political tension, the final assembly would be conducted in the U.S. by S7's subsidiary, Sea Launch Limited. Another option for the Sea Launch is the new development of the JSC SRC Progress—the Soyuz-5 rocket, which could be very useful and capable of fulfilling the S7 goals.

Space International Services

Space International Services (SIS) is a joint venture between Russian and Ukrainian partners that was created to support the exploitation of Zenit (which comprises roughly 70% Russian and 30% Ukrainian components) by Sea Launch under its Land Launch programme at Baikonur.

Contracts for the launch services from Baikonur were implemented via a Sea Launch subcontract with SIS. More specifically, Sea Launch maintained the functions of mission management, quality assurance and hardware acceptance procedures, while SIS provided programme coordination and control including the provision of all launch system components, mission integration and launch operations of Zenit.

Following its recovery from bankruptcy, Sea Launch carried out the first Zenit land launch from Baikonur in 2011, but in the same year it put in question the future of its land launch operations. During the same period, SIS started to increase its efforts in the commercialisation of Zenit, advertising its marketing of the Zenit launch services from Kazakhstan. In part as a result of the new offer by SIS, which Sea Launch regarded to be more competitive than its own, and in part owing to emerging difficulties caused by the fierce competition with SpaceX, in 2012 the company took the decision to shift its focus exclusively on maritime platform solutions. Accordingly, in 2013 Land Launch services were removed from its advertised offer. On the same year, SIS performed its first a Zenit-3 SLB launch from Baikonur. However, the 2014 political crisis between Russia and the Ukraine (see in Chap. 3) led to a suspension of Zenit operations (with the exception of a launch at the end of 2015). Owing to the difficulties incurred by the Zenit programme and the recent acquisition of Sea Launch (which has announced its interest to re-include Zenit-3M launch from Baikonur in its portfolio possibly without SIS subcontract), it is unlikely that SIS will maintain significant activities in the coming years.

International Launch Services

International Launch Services (ILS) was originally formed in 1995 as a private spaceflight partnership between Lockheed Martin, Khrunichev and Energia to co-market commercial launches on the U.S. Atlas and Russian Proton launch vehicles.

While the company has been extremely successful in the international launch market during the 2000s, in 2006 Lockheed Martin announced the intention withdraw from the JV and to sell its share in ILS to Space Transport Inc., a new entity that was constituted specifically for this transaction. Following the 2008 transfer of ownership, Khrunichev acquired all of Space Transport's interests and became the majority shareholder, holding 83% of the shares while being in control of almost 70% of Proton production. The 2014 incorporation of Khrunichev and Energia into URSC and the 2016 incorporation of URSC into Roscosmos SC now fully place ILS under the state control of Roscosmos. Nevertheless, as of December 2017, the company remains a U.S. company with its headquarters in Virginia.

While the commercial launch capacity of ILS has been at around 7–8 annual launches per-year during the 2000s, in the past ten years Proton vehicles have been prone to several failures, thus affecting the commercial position of the launch provider. Because ILS traditional customers started to look for alternative launch solutions, ILS inevitably saw its launch rate projections reducing from the previous 7–8 launches per year to 3–4 launches. By 2015 the company also had to cut its workforce by 25%. Furthermore, the fundamental engine problems that emerged in 2015–2016 came at a moment where the company had just started to recover from its difficult situation in the commercial market. The unsolved problems on the lower stages and the possible schedule delays imposed by engine reworks may now compromise the customer confidence in Proton M. In light of this unpleasing background, in 2015, ILS partially re-oriented its strategy by beginning to offer the Angara launch vehicle for commercial customers. The first commercial launch is currently planned in 2020.

International Space Company

The International Space Company (ISC), Kosmotras is a joint venture created in 1997 with the participation of Russian and Ukrainian space agencies and their respective industrial companies for the exploitation of the Dnepr rocket. Unlike other joint ventures, the launcher system was of mixed origin, with Ukrainian partners being responsible for design supervision over the Russian converted ICBM and providing the payload adaptor.

In early 2011, following a 2–year negotiation process, Kazakhstan became a minority shareholder (10%) of the JV, represented by Kazakhstan Garysh Sapary (KGS), a joint stock company established in 2005 under the control of the Kazakh government. The remaining 90% was split equally between Russian and the Ukrainian stakeholders. Subsequently, in 2012 the government of Astana made a request to increase its participation in Kosmotras to 33.3% in order to achieve parity with the Russian and Ukrainian partners. The request was approved by the Russian Federal Anti-Monopoly Service in February 2013. However, this rearrangement did not see the light as according to the Kazakh space agency the deal was not supported by former management of Roscosmos. Following the Russia-Ukraine tensions of 2014, the Dnepr programme was suspended in early 2015 and Kazakhstan decided to withhold the stock acquisition process pending receipt of an official document from the Russian government confirming the continuation of the Dnepr conversion programme.

In the meanwhile, all remaining Ukrainian interests as well as the interests previously held by the Russian Askond company were acquired by Russian billionaire Sergey Nedoroslev of the Kaskol Group and a team of investors in the second half of 2015. The repatriation of ISC's interests in Russia should allow to restart Dnepr launch activities. However, no signs to this development are visible as of December 2017.

Eurockot Launch Services

Eurockot Launch Services was created in 1995 to commercialize the Rockot launch system for international operators of LEO satellites. It is a joint venture of EADS Astrium and Khrunichev State Space Research and Production Space Center (KhSC), holding 51 and 49% respectively. Dedicated launch facilities have been established in Plesetsk Cosmodrome in Northern Russia.

Building on Russian military heritage technology, Eurockot has positioned itself as a cost-competitive launch service provider. It has been offering Rockot launch services to international customers, while Khrunichev has continued to offer launch services to domestic institutional customers. However, following a 2017 decision of Khrunichev to discontinue the production, Rockot is to be retired following the two remaining launches, which have been earmarked for Sentinel missions.

Annex E
Past and Future Europe–Russia Space Cooperation Timeline

Year	Items
1991	Signature of ESA-Roscosmos Framework Agreement on Cooperation
1994	European astronaut's mission on MIR space station
1995	Establishment of an ESA Representation Office in Moscow Creation of Eurockot Launch Services
1996	Creation of Starsem Joint Venture
1997	Signature of EUMETSAT-ROSHYDROMET Framework Agreement on Cooperation
2001	First European astronaut mission to the ISS aboard Soyuz-TM33 Signature of ESA-EC-Roscosmos Joint MoU on "New Opportunities for a Euro-Russian Space Partnership"
2002	Establishment of an "EU-Russia Dialogue on Space Cooperation"
2003	Joint Workshop to Promote EU-Russia Space Cooperation Signature of ESA-Roscosmos Intergovernmental Agreement on Cooperation and Partnership
2005	Adoption of the EU-Russia Roadmap to the Common Economic Space
2008	Launch of the first ATV to the ISS
2011	First Launch of Soyuz from the Guiana Space Centre
2013	Signature of ESA-Roscosmos Cooperation Agreement on the Exploration of Mars and other Bodies in the Solar Sysems Signature of MoU on the definition of future ESA-Roscosmos cooperation for the robotic exploration of the Moon
2016	Launch of ExoMars Orbiter
2017	Signature of ESA-Roscosmos Cooperation Agreement on the Exploration of the Moon
2018	*Launch of Bepi-Colombo* *Launch of MLM with European Robotic Arm to the ISS*

(continued)

© Springer International Publishing AG, part of Springer Nature 2019
M. Aliberti and K. Lisitsyna, *Russia's Posture in Space*, Studies in Space Policy 18,
https://doi.org/10.1007/978-3-319-90554-9

(continued)

Year	Items
2019	*Launch of Luna-Glob Lander*
2020	*Launch of ExoMars Lander*
2021	*Launch of Luna-Resurs*
2022	*Launch of JUICE*
2024	*Launch of Phobos-Grunt-2*
2025 +	*Launch of Luna-Grunt/LPSR*

Printed in the United States
By Bookmasters